SpringerBriefs in Applied Sciences and Technology

SpringerBriefs present concise summaries of cutting-edge research and practical applications across a wide spectrum of fields. Featuring compact volumes of 50–125 pages, the series covers a range of content from professional to academic.

Typical publications can be:

- A timely report of state-of-the art methods
- An introduction to or a manual for the application of mathematical or computer techniques
- A bridge between new research results, as published in journal articles
- A snapshot of a hot or emerging topic
- An in-depth case study
- A presentation of core concepts that students must understand in order to make independent contributions

SpringerBriefs are characterized by fast, global electronic dissemination, standard publishing contracts, standardized manuscript preparation and formatting guidelines, and expedited production schedules.

On the one hand, **SpringerBriefs in Applied Sciences and Technology** are devoted to the publication of fundamentals and applications within the different classical engineering disciplines as well as in interdisciplinary fields that recently emerged between these areas. On the other hand, as the boundary separating fundamental research and applied technology is more and more dissolving, this series is particularly open to trans-disciplinary topics between fundamental science and engineering.

Indexed by EI-Compendex, SCOPUS and Springerlink.

More information about this series at http://www.springer.com/series/8884

Dinesh Kant Kumar · Peterjohn Radcliffe

Teaching Surrounded by Smart Phones

Springer

Dinesh Kant Kumar
Electrical and Biomedical Engineering
RMIT University
Melbourne, VIC, Australia

Peterjohn Radcliffe
Department of Engineering
RMIT University
Melbourne, VIC, Australia

ISSN 2191-530X ISSN 2191-5318 (electronic)
SpringerBriefs in Applied Sciences and Technology
ISBN 978-981-15-1400-5 ISBN 978-981-15-1401-2 (eBook)
https://doi.org/10.1007/978-981-15-1401-2

Thanks to Mr. Susmit Bhowmic for his artwork.

This Springer imprint is published by the registered company Springer Nature Singapore Pte Ltd.
The registered company address is: 152 Beach Road, #21-01/04 Gateway East, Singapore 189721, Singapore

The new classroom

Contents

1 About Smartphones and Education 1
References 6

2 How and Why Students Use Smartphones 7
2.1 Some Key Apps to Understand 13
2.2 What Are the Effects of Smartphones on People? 21
References 25

3 The Effect of Smartphones on Traditional Education 27
3.1 Access and Equity 34
References 36

4 Using Smartphone to Enhance Education 39
4.1 Some Useful Tools 40
4.2 Tools for Policing the Use of Smartphones 44
4.3 Methods and Techniques 47
References 51

5 Measuring the Effects 53
References 56

6 Case Studies 57
6.1 Case Study 1: Policing the Use of Smartphones—Large Classrooms 58
6.2 Case Study 2: Policing the Use of Smartphones—Small Classrooms 60
6.3 Case Study 3: Sharing the Teaching Material On-Line in the Class 61
6.4 Case Study 4: Using Purpose Developed on-Line Material 64
6.5 Case Study 5: Using Clickers in Class 66

6.6 Case Study 6: The Flipped Classroom and Smartphones 68
6.7 Case Study 7: Using Kahoot in Class 71
6.8 Case Study 8: Google Forms 73
6.9 Case Study 9 75
References 77

7 Successful Strategies 79
Bibliography 83

About the Authors

Prof. Dinesh Kant Kumar and Dr. Peterjohn Radcliffe Each of us were being challenged in our classes due to regular disturbances caused by smartphones. When we spoke with other colleagues, we realised that the problem was widespread, and no one seemed to have a solution. Between the two of us, we have over 50 years of teaching experience, and realised that we were lost with this sudden growth in smartphone technology. We also realised that this was leading to significant learning issues for the students. This book is our effort to share our research and ideas in understanding, mitigating and harnessing smartphones in our own classrooms.

Prof. Dinesh Kant Kumar, B.Tech. from IIT Madras, and Ph.D. from IIT Delhi, is a Professor at RMIT University, Melbourne, Australia. He has published over 400 papers, authored 5 books and is on a range of Australian and international committees for Biomedical Engineering. His passion is for affordable- diagnostics and making a difference for his students. His work has been cited over 5600 times and he has also had multiple successes with technology translation.

Dr. Peterjohn Radcliffe, B.Eng. from Melbourne University and Ph.D. from RMIT, is a senior lecturer at RMIT University in Melbourne Australia. He has been teaching since 1986 and maintains strong links with industry both with project work and consulting. His passion is ensuring that RMIT graduates have both the professional and technical skills to be keenly sought after by industry. His technical interests include networking, network security, IoT, embedded systems and mobile computing.

Chapter 1
About Smartphones and Education

This book describes our research in the area of teaching University students when many of them carry smartphones and may often be distracted by these devices. The aim of this chapter is to provide context to the problem of smartphones in education. These devices are everywhere, in classrooms, and easily sneaked even in the examinations. Smartphones are directly affecting student achievements, for example by distracting them, and affecting behaviour such as increasing expectations for instant results. It is essential that the teachers and academics are aware of the issues and evolve their teaching styles to cope with the reality of education while surrounded by smartphones.

D. K. Kumar and P. Radcliffe, *Teaching Surrounded by Smart Phones*,
SpringerBriefs in Applied Sciences and Technology,
https://doi.org/10.1007/978-981-15-1401-2_1

Smartphones even sneak into exams

Introduction

Simply speaking, smartphones have changed humanity. These devices that were not known two decades ago, now dominate our world and have the amazing capabilities that have changed the way we socialize, spend our private time, learn, and the way we work. These devices give us the capacity to obtain information from diverse sources, play games or socialize while walking in a garden or sitting in a meeting. Many of us have become highly dependent on these devices and would not leave our homes without them.

Smartphones come in a range of sizes, and pricing, but the fundamental look and functionality are similar. While top-end devices can be expensive, there are many very inexpensive devices with excellent capabilities. The full content of the Internet and its many services is now available anywhere, anytime, to nearly anyone. Facebook, twitter, dating apps, email, video chat, movies and more are available to everyone at a nominal cost. These things are so attractive they have even been blamed for a reduction in the frequency of sex for young couples (Roe 2017). Smartphone usage varies immensely but according to Deloitte ("Mobile Consumer Survey" 2018): typically, 70% of 18–24-year-old admit they use their phones excessively, 50% are attempting to limit their usage, but only one in five of those are succeeding. Additionally, young people underestimate their phone usage typically by a factor of two (Gregoire

2015). Psychologists are treating people with excessive attachment to smartphones, referred to as smartphone addiction (Kim 2013; Lin et al. 2016) and scientists are investigating this as a pathological condition (Lee et al. 2014).

From literature and popular media, it is evident that the smartphone is causing major behavioural changes particularly in young people. It is this cohort who dominate the education system and to presume that their education is unaffected by this revolution is both naive and runs the risk of producing steadily worse education outcomes. While most of the students were born after internet and mobile phones were available, the academics were born in a previous era and may not be able to understand the compelling factors that are faced by the young students.

There are a number of organisations around the globe that have attempted to ban the presence of smartphones and other Internet devices in the classrooms (McKinnell and Tiller 2018). Exam scores for lower performing students have been found to rise when phone bans were imposed (Beland and Murphy 2016). An attempt to quantify the problem conducted in France and the UK identified that in high schools, students on average access Internet more than two times every hour when phones are banned. They also found that students can spend over 20% of their classroom time engaging with the Internet. Thus, begins the debate as to whether smartphones can be used to enhance the student learning activities. The pros for the use of digital devices in the classrooms (Anshari et al. 2017, Kemp et al. 2014) are:

1. Using educational learning apps: The apps can help them record material, improve time management and organization skills.
2. Incorporating digital platforms into lessons: Social media can encourage student participation in the discussion, especially helpful for introverted students.
3. Supplementary digital materials and extra information: Teachers can provide students with extra resources.

The Cons for this are:

1. Distractions: Use of cell phones in the class permits them to check social media and text friends leading to distractions for them and their nearby peers.
2. Cyberbullying: It is now recognised that cyberbullying is very much present in the lives of young people. Easy access to smartphones can make it harder to manage this problem.
3. Cheating: smartphones may pose further difficulties for the management and minimization of this widespread problem. It is easier for students to cheat because the smartphone can be preloaded with material or used to access help and material from outside.
4. Disconnection from face-to-face activities: Heavy use of non-social media can disconnect students from other students, the education activity, and spiral them into isolation.

Given these problems, many teachers and academics have attempted to ban smartphones in their classrooms. In July of 2018, the French government banned cell phones in schools, from kinder through to 9th grade ("France bans smartphones in schools" 2018). Some primary and secondary schools in the USA have also enacted similar policies and some Universities also attempt to ban the use of smartphones in the classroom. However, as Anita Charles, Director at Bates College remarks ("Why banning cellphones" 2018) that such policies are largely unenforceable. Teachers and students find ways to subvert these rules and regulations, but the net result is usually just hiding the problem rather than solving the problem. In a university setting the banning of smartphones is largely impossible, except in exam style conditions.

Some of us may wish smartphones did not exist in our classrooms, but the reality is that they are here to stay, particularly in the university context. An understanding of how students use their smartphone and how that affects their motivation and behaviour may lead to improved educational outcomes. The intent of this book is to discuss strategies that can help the academics and teachers identify methods that can help them improve the learning outcomes of their students.

About the Book

This book is the result of our realisation of this issue, and recent research in this field, and it is expected to improve the classroom outcomes in the modern era. It is essential to realise that the Internet and smartphones are a very recent phenomenon and are evolving very rapidly. New devices, software apps, and methodologies are emerging on a regular basis and may upset previous understandings. A major rationale for this book is to share different perspectives at the current time and to encourage continuing dialog so the education community can react in a timely manner to any new developments.

The purpose of our work was to understand the problem, study methods that people are using to mitigate the negative effects of smartphones, and determine which methods are the most suitable for our students. In the process we realised there are significant differences in the methodologies and educational environment faced by academics and teachers. We have examined real classroom experiences to identified suitable case-studies to help understand the problem and the different proposed solutions. In this book, we report some of the relevant case-studies that may facilitate academics in different circumstances to determine the methodologies most suitable for them. We would however like to point out that these are only examples, and many academics would have their own situations, strategies and preferences. To complicate matters even further new devices and software are constantly being developed which means that this entire field is dynamic and changing. What is an excellent strategy today may be counterproductive give the next popular smartphone app. One of our main hopes is that this book will help start an ongoing conversation about these issues.

The first step in coping with the smartphone revolution is to understand what services are available and how this affects student behaviour, their value system, and what they see as important. Chapter 2 examines these issues and also shows that the effects are not homogeneous, and the relationship is complex; smartphones can affect different people in different ways even within a group of seemingly similar people.

Chapter 3 examines what can happen to traditional education techniques when smartphones enter the teaching environment. We demonstrate that the educational outcome can actually be made worse by misguided attempts to enhance student's digital experience while at university. This chapter shares a case study to demonstrate the results when the digital devices were introduced in teaching but without careful planning. If smartphones are to be used in the classroom for active teaching, there are issues of equity and access for all the students that must be considered. Some phones have far greater capabilities than others and may also have greater access to Internet services. Chapter 3 also looks at these issues and suggests methods to manage these issues. It discusses the minimal range of services and features that are universal from the very cheap to the very expensive smartphones, and how to ensure equity and access to all the students.

Chapter 4 looks at a variety of methods and tools which can either limit the effects of smartphones or alternatively co-opt them into the education process. Some methods will suit your situation, others will not. This chapter discusses some example case studies to identify the strengths and weaknesses of different methods and styles. There is no one panacea that suits every situation and the educator must make a choice about what to methods try based on the individual characteristics of the classroom in question.

Education by "what feels right" is not scientific and can result in horrendous mistakes such as the rejection of phonetic reading debacle in Australia, the UK and the USA (Rowe 2005; Rose 2006). To help avoid such problems, Chap. 5 discusses methods by which the effectiveness of smartphone methods can be measured.

Chapter 6 provides a variety of case studies which illustrate both success and failures as education innovations attempt to cope with smartphones. What works or does not work is dependent on the background and culture of each class and so this background is also detailed.

Chapter 7 provides a summary of strategies that have been investigated by the authors in a table form to provide a quick reference to key solutions.

Smartphones are certainly here to stay, and they can certainly affect learning outcomes. We educators must, at the minimum, analyse how these devices will affect our particular teaching activities. There is no common effect on all students nor common solutions to all problems. In some cases, few changes need be made to the education process. For example, the training of medical students does not seem to be negatively affected by smartphones. If anything, the constant ability to view the web and educational resources has helped medical education. Such students come from the top of the high school cohort, are highly motivated, and are often in a competitive situation to get into later years, minimize fees, or obtain scholarships. At the other extreme we have seen demotivated students in education programs they did not like. Furthermore, there were no consequences for poor performance and so these students just played with their mobile phones during compulsory attendance lecture or tutorials. Your educational situation is likely to sit in between these two extremes and you will need to consider both your students and the nature of smartphones before deciding how to modify your teaching practices.

References

Anshari, M., Almunawar, M. N., Shahrill, M., Wicaksono, D. K., & Huda, M. (2017). Smartphones usage in the classrooms: Learning aid or interference? *Education and Information Technologies, 22*(6), 3063–3079. https://doi.org/10.1007/s10639-017-9572-7.

Beland, L. P., & Murphy, R. (2016). Ill communication: technology, distraction & student performance. *Labour Economics, 41*, 61–76.

France bans smartphones in school - The Washington Post. (2019). Retrieved June 24, 2019, from https://www.washingtonpost.com/technology/2018/07/31/france-bans-smartphonesschool/?noredirect=on&utm_term=.12f04620af74.

Gregoire, C. (2015, November 3). You Probably Use Your Smartphone Way More Than You Think. *Huffington Post*. Retrieved from https://www.huffpost.com/entry/smartphone-usage-estimates_n_5637687de4b063179912dc96.

Kemp, A., Preston, J., Page, C., Harper, R., Dillard, B., Flynn, J., et al. (2014). Technology and teaching: A conversation among faculty regarding the pros and cons of technology. *The Qualitative Report, 19*(3), 1–23.

Kim, H. (2013). Exercise rehabilitation for smartphone addiction. *Journal of Exercise Rehabilitation, 9*(6), 500–505. https://doi.org/10.12965/jer.130080.

Lee, H., Ahn, H., Choi, S., & Choi, W. (2014). The SAMS: smartphone addiction management system and verification. *Journal of Medical Systems, 38*(1), 1. https://doi.org/10.1007/s10916-013-0001-1.

Lin, Y. H., Lin, Y. C., Lee, Y. H., Lin, P. H., Lin, S. H., Chang, L. R., et al. (2015). Time distortion associated with smartphone addiction: Identifying smartphone addiction via a mobile application (App). *Journal of Psychiatric Research, 65*, 139–145.

McKinnell, J., & Tiller, S. (2018, December 13). Mobile phones will be banned in NSW primary schools from next year [Text]. Retrieved June 24, 2019, from ABC News website: https://www.abc.net.au/news/2018-12-13/nsw-phone-ban-aims-to-reduce-bullying/10612950.

Mobile Consumer Survey. (2018). Deloitte Australia. Technology, Media & Telecommunications, Mobile, Trends. (n.d.). Retrieved June 24, 2019, from https://www2.deloitte.com/au/mobile-consumer-survey.

Roe, D. (2017, September 21). Don't Worry If You're Not Having Sex—Apparently, No One Is. Retrieved June 24, 2019, from Men's Health website: https://www.menshealth.com/sex-women/a19535722/why-are-young-people-not-having-sex/.

Rose, J. (2006). Independent review of the teaching of early reading: Final report. Retrieved June 24, 2019, from http://dera.ioe.ac.uk/5551/3/5d970d28fc535dc54eb4dee9995bef36.pdf.

Rowe, K. (2005). National inquiry into the teaching of literacy (Australia), &. teaching reading: Report and recommendations. *Teaching and Learning and Leadership*. Retrieved from https://research.acer.edu.au/tll_misc/5.

Why banning cellphones in schools misses the point. (2018, March 23). Retrieved June 24, 2019, from News website: https://www.bates.edu/news/2018/03/23/why-banning-cellphones-in-schools-misses-the-point/.

Chapter 2
How and Why Students Use Smartphones

This chapter contends that smartphones have a significant effect on the behaviour and values of people, young people in particular. It is further contended that it is not possible to design an effective education system without an understanding of these effects. The bulk of the section looks at a variety of services and applications and how this affects behaviour and values of smartphone users. It is essential that the strategies being considered by the teachers to improve student learning are based on this information.

Smartphones have brought about dramatic changes particularly in the younger population. According to Deloitte some 60% of young people think they use their smartphones too much ("Mobile Consumer Survey" 2018). It is impossible to understand how the smartphone affects the education process without understanding why people, particularly young people, use smartphones. The services available have clearly affected the behaviour of young people and what they value.

D. K. Kumar and P. Radcliffe, *Teaching Surrounded by Smart Phones*,
SpringerBriefs in Applied Sciences and Technology,
https://doi.org/10.1007/978-981-15-1401-2_2

Smartphone use is starting earlier and earlier!

As a starting point for understanding the problem, it is essential to observe what people do with their smartphones in their normal lives. People use them on public transport and in waiting areas such as doctor's waiting rooms, during an official meeting, when walking with friends, when sitting with family, or even when having dinner with friends. In education smartphone use can be seen in the classrooms, during lectures, and while listening to online lectures in their own homes. Young people have grown up surrounded by these devices and do not know a world without them. Even some of the older cohort make heavy use of smartphones.

It is essential to realise that smartphones are not evil, and many of the good things we enjoy are thanks to these little devices. Before free tools such as Google Maps started in 2004, how did we drive and reach our destination without the help of the GPS on smartphones? How did we meet people without these magical devices? And, how did we check the facts from fiction while listening to a seminar or speech? We cannot imagine the world without these, and the youth have not known a world without these facilities.

Smartphones present great advantages, and dangers

The educational environments for many of the students is not dissimilar to attending official or personal meetings; they have to wait for activities to start, or for something which will engage their interest. Every university student, young or old, can remember lectures where nothing useful was said. While waiting for things to happen, is it not natural for most of us to reach out to the smartphone and check our messages, or just read the latest news?

When observing smartphone use, care must be taken first not to be seen as intrusive, and secondly not to be intrusive or to invade peoples' privacy. A casual glance may be appropriate but not careful observation which, besides being unethical, could bias the results. In an educational setting you might be able to see what students are doing on their phones but normally the sight of an academic nearby will result in the phone being put down. You may also ask the tutors to tell you what students are doing on their smartphones. Tutors do not seem to excite the same stress as does an older academic. Public transport is also a good option to observe smartphone usage because one can observe the screens of people while casually walking along a rail carriage or bus aisle. From our observations, the majority of smartphone usage appears to be filling in time while nothing else is happening. A smaller group will continue to use their smartphone when they should be doing something else.

Our observation of smartphone use in public has observed the following apps being used; Facebook is very popular, SMS both for work or as a social activity, games, YouTube, news web sites, music, email for work or private purposes, gambling, or on rare occasions people actually use their smartphone as a phone. While this list is not exhaustive, and there are many other Internet services that are used, and the list is growing, these are the ones that are currently very common. Such activities must reduce a person's ability to cope with boredom or to wait for more interesting things such as the section of the lecture that is considered as useful. The availability of constant stimulation via smartphones can lead to a form of addiction (Kim 2013), which we have personally seen at the university level. Addicted students will choose stimulating but in the long-term useless activities over less stimulating but important activities as judged in the longer term of days or months. We are not aware of any university that has policies for detecting and offering help to students addicted to their smartphones. This is a difficult policy area but an important one as it may well improve retention rates and help more people gain a worthwhile career.

Waiting areas are also interesting to watch though it can be harder to see for what purpose people are using their phones. We have observed one patient in a doctor's waiting room ask if they could delay seeing the doctor because they were reading breaking news on their smartphone. Many people have missed their train or bus station because they were engaged with an item on their smartphones. Short-term, attention grabbing items on the smartphone may not be the most important task at hand but because they are attention grabbing, they often get priority over more important but less exciting tasks.

Smartphones have transformed even the doctor's waiting room

There is a further effect than the use of Internet services just filling in the time of the users. Many users suffer from the "down the rabbit hole" effect due to the smartphones, which result in a small search resulting in a few hours of distraction. There may be an interesting post on Facebook that leads to another interesting thing, or reading in-depth about a topic, or searching for other opinions. Great swaths of time can disappear at high levels of concentration. If this happens in your class, then the student will hear little of what you say.

SMS or similar apps such as WhatsApp are a very commonly used service on smartphones that can send and receive short text messages between individuals and groups of people. Usually, these people know each other, and SMS can be seen as a remote conversation though it also may take place within a classroom or in a meeting. This may be considered to be a sub-meeting taking place inside a meeting or lecture. SMS is an old technology with a lot of competition from other technologies, but it continues to be heavily used. Our students use it to convey information but there appears to be a strong social interaction facet to the SMS exchanges and so most SMS conversations appear as socializing at a distance. There appears to be differences in the usage of these based on gender and age (Smythe and Costall 2003). It has been reported that females appear to use SMS more for maintaining and strengthening relationships whereas males tend to use it more for practical purposes such as conveying facts (Geser 2006).

Another very popular use of smartphones is based on the photo and video imaging capability of these devices. An observation based on the advertisements of smartphones show that the image quality and ability to take selfies are very important features. Social media such as Instagram, Facebook, and WhatsApp thrive on images and young people are heavy users of these services.

The great selfie, look at me!

The cost of taking and storing images has become very small, and most phones are now equipped with cameras and enough memory to take large number of high-quality images. It is not uncommon to see students taking photos of the screen in the class, or even recording a video of the lecture. We have observed students in a lab situation building hardware. When the first group got their hardware working, everyone else took a photo with their smartphones with the aim of duplicating the first groups success.

Another successful way to find out how your students use their smartphones is simply to engage them in conversation. Many students are happy to discuss their favourite apps and websites and how much they use them. Like all people, students will want to maintain a certain public image and so will omit things that go against that public image. For example, do not expect students to tell if they use gambling sites or apps, and they may underquote just how much time they spend on their smartphone. In our experience, Facebook and other social media apps are the most popular, followed by email, and YouTube. Your class may well be different so try to start a dialog with your students to find out what they use, and why they use it.

2.1 Some Key Apps to Understand

There are a huge variety of apps available for smartphones, furthermore an app such as Facebook or Twitter can be used for a remarkable variety of things from simple time filling, real learning, to actively running a business, and illegal activities. It is important to be aware of the most common apps and their basic functionality. Treat this chapter as a checklist and read in-depth any app which you have not used.

The designers of smartphone apps have delved deep into the research on gambling and human behaviour to make their products irresistible (Parkin 2018). As noted in the New York Times, "Tech companies understand what causes dopamine surges in the brain, and they lace their products with 'hijacking techniques' that lure us in and create 'compulsion loops'." (Brooks 2017). Apps such as Facebook use similar strategies to those used in gambling and so have become one of the most used apps in the world.

Facebook

Facebook is a global reach social media company that has changed the way many people in the world interact. This simple concept allows people to share their 'lives' with friends. Users on Facebook create their 'profile', which provides the visitor some information about themselves and a 'wall' on which people can post their comments. The users with a profile can post their stories onto Facebook and make 'friends'. 'Friends' can see each other's profiles and share stories, photos, video, events, and other content with them. Somewhat addictively new posts cause a "ping" on the smartphone which is a call for your attention. Some have likened this to a Pavlovian response which trains the Facebook user to abandon anything else they are doing and use Facebook.

Email

This is perhaps the first major online communication tool, which allows the user to deliver their mail to the other person instantaneously and essentially for free. Email has changed our behaviour such that we have stopped receiving the regular paper mail, and most professional, legal, and other communication is now by email. Most of the providers of the email would allow the users a free style of writing their messages. Sociologists believe that the instantaneous nature of the emails has made us very impatient (Farmer 2003; Tyler and Tang 2003). Some people, including our own students, have started to expect an instantaneous the response to their messages and as a result, the time devoted to other activities, such as studies, reading or personal lives have been eroded.

Twitter

Twitter is an American online news and social networking service on which users post and interact with messages known as "tweets". Tweets were originally restricted to 140 characters, but on November 7, 2017, this limit was doubled to 280. While emails have gone to the point of the receiver being expected to respond nearly instantaneously, Twitter has brought it to the next level. Often this is used by celebrities or inspiring to be celebrities, with leaders and followers, and topics may range from politics to fashion, or anything else, including academics. President Trump of the USA is noted for using Twitter to announce policy and keep in the news media spotlight.

Youtube

YouTube is video-sharing website headquartered in California. Founded by Chad Hurley, Steve Chen, and Jawed Karim in February 2005, this allows the user to upload videos on the server which can then be viewed by anyone who has access to the Internet. The supplier of the video can earn some money based on the number of viewers, and the contents range from entertainment, to educational lectures, lectures on current social and political topics. YouTube is now owned by Alphabet (Google) and is driven by their amazing personal data mining capabilities. Any video that gets played also advertises a range of other videos based on your personal preferences. Many people find YouTube provides a down-the-rabbit-hole experience where one video leads to another, and another, and suddenly the user realizes that hours have passed by.

Instagram

Instagram is a photo and video-sharing social networking service owned by Facebook, Inc. It was created by Kevin Systrom and Mike Krieger and launched in October 2010 exclusively on iOS but is now available on many platforms. This is popular with teenagers. It has even removed the need for writing or reading as a story can be built out of video and audio clips and images. While it is often used to communicate with friends, it is also used for planning activities after-office or University hours.

Snapchat

Snapchat is a multimedia messaging app used globally, created by Evan Spiegel, Bobby Murphy, and Reggie Brown, former students at Stanford University, and developed by Snap Inc., since 2011. Snapchat is a popular messaging app that lets users exchange pictures and videos (called snaps) that are meant to disappear after they're viewed. It's advertised as a "new type of camera" because the essential function is to take a picture or video, add filters, lenses or other effects and share them with friends. The philosophy appears to be giving the other person access to one's personal look. While generally harmless, it has been commonly associated with issues such as sexting.

LinkedIn

LinkedIn is a business and employment-oriented service that operates via websites and mobile apps. Founded on December 28, 2002, and launched on May 5, 2003, it is mainly used for professional networking, including employers posting jobs and job seekers posting their CVs. This site is very popular among professionals seeking employment or even to get noticed by their professional peers. It could be very popular with students in their final year of University, or among the post-graduate students.

ResearchGate

ResearchGate is a social networking site from Berlin, Germany (2008) for scientists and researchers to share papers, ask and answer questions, and find collaborators. This allows researchers to perform their research on this site, and while it is interactive, it does not require a spontaneous response. It is unlikely to attract students in the start of their University, though it may be popular among research students, or students undertaking challenging projects.

Academia

"Academia.edu" is a commercial social networking website for academics. The platform can be used to share papers, monitor their impact, and follow the research in a particular field. It was launched in September 2008, with 39 million unique visitors per month as of January 2019 and over 21 million uploaded texts. Similar to ResearchGate, it is unlikely to attract the young students, and students using this are generally going to be the ones conducting research and literature review.

Reddit

Reddit is an American social news aggregation, web content rating, and discussion website. Registered members submit content to the site such as links, text posts, and images, which are then voted up or down by other members. This has an interactive element, which makes it very engaging for the members, and the topics are often very current news items. Thus, the time is of essence and users are keen to scan the site frequently and respond quickly. It is quite possible that this site may result in the students being significantly distracted. There is low chance of this being used for academic purposes.

Shopping Apps

Shopping is another common use of smartphones and other Internet connected devices, and recent studies revealed that some young people had not shopped in physical stores for over a year. Shopping online has become extremely common for many of us, and it is not uncommon to be distracted by shopping messages, especially when these are offering discounts that are valid for only a short time. Then there are auction houses that attract people for online bidding or bargaining for a product or even experiences such as for travel. Some of the commonly used sites are: eBay, Amazon, LightBox, Alibaba, SnapDeal, and then there are airlines, and travel agencies. While the conventional shopping sites are unlikely to distract students during their learning activities, the ones that are interactive such as auctions are quite likely to be very distracting.

Smartphones have transformed shopping

Netflix

Netflix, Inc. is a USA based company that provides video on demand service which was founded in 1997. While this is not a new company or concept Netflix has become very popular due to the availability of high-speed wireless Internet, the ability of mobile devices to display high-quality videos, and the significantly reduced cost of membership of Netflix. Thus, it is not uncommon for a student sitting anywhere, including the classroom, to attempt dual tasking, viewing a movie or program while also listening to the lecture. This is not an interactive site, and all the movies are available on-demand, which means that the viewer can delay their watching on the off chance that the lecturer has something interesting to say.

Messaging sites: Whatsapp/Viber/Emo/Tango/Messenger/Skype/WeChat

There are a large number of sites and apps for people to send messages. While most of these are very similar, each of these have their own appeal and audience. In general, they allow users to exchange messages between 2 people, or in a group. Some of these sites or apps are for posting messages to be shared in a small community while there are others that are for general public. Many of these also provide voice-over IP service that is used for telephone calls, or even video calls, and are an extremely popular and free substitute for long-distance phone calls. With the availability of free wireless Internet in most institutions, it is not uncommon for people to transmit live their activities to friends. This may also be used by students or the teachers to reach the students who are unable to come to the class.

Dating, Flirting and Relationship Sites

Shaadi/OKCupid

Dating sites are extremely popular, especially among the young. These cater to all kinds of markets, from the traditional, doing the role of what the local newspapers did many years ago, to the very specific such as for very short-term relationships. Sites such as "Shaadi" are more traditional and are targeting the matrimonial market. Typically, these are unlikely to be demanding spontaneous responses from the users. However, there are others that are designed for more casual relationships such as OKCupid, which is an internationally operating online dating, friendship, and social networking website that features multiple-choice questions in order to match members. Active since 2004, it attracts paid membership but is also supported by advertisements for unpaid users.

Tinder/Bumble/Coffeemeetsbagle/wine/Tumblr

Tinder is a location-based social search mobile app, most often used as a dating site, that allows users to like or dislike other users, and allows users to chat if both parties like each other. Since 2012). There are many such sites, and while some are free to the users, often these charge the users for the services. These allow the users to advertise their personal profile and photos online—which may be to a select audience or to anyone. Such sites are usually used to make casual friends and dates though rarely for long-term relationships. Typically, these are designed for spontaneity and the active users are more likely to be distracted by these.

Love will never be the same after smartphones

Online Gaming

Computer based gaming has been a major business since the early times of the computers. With improved graphics, Internet connectively and mobile devices, computer games have evolved and now contain sophisticated features such as artificial intelligence and amazing graphics and audio. Some apps contain simulations that are very realistic such as FIFA, Minecraft, and Call of Duty. The game Crossfire claims 660 million registered users. Some games have significant educational value and are commonly employed for training people to drive, or fly, and modified versions are here are some for training doctors and surgeons. Games often have intentions other than earning money for the game's creators. Violent shooting games have the support of the US military (Romaniuk and Burgers 2019), some encourage gambling, or support product franchises.

The most popular online games are mostly combat focused and in 2019 include Fortnite (a battle royale), PlayerUnknown's Battlegrounds, League of Legends, and Hearthstone.

Kongregate

Kongregate is an American mobile, PC, and Console publisher and web gaming portal. The website features over 110,000 online games and 30+ mobile games available to the public.

Ballpool

Another interactive online game that enables users to play billiards online, and in real-time.

Beteasy, SportsBet/Bet 365

BetEasy is an Australian corporate bookmaker offering online sports and racing betting platforms since 2014. Sportseasy is a gambling site that offers Apps for smartphones. Research has found that Internet gambling is widespread and highly addictive (Wardle et al. 2011; Ng and Wiemer-Hastings 2005). While the gambling sites that are based on casino type games are not time-bound, the ones with real time activities such as sports require attention at particular times may interfere with classroom activities.

Phone betting is now a huge industry and occupies many people's time

OnlineCasino

Online gambling has grown very quickly and while many countries are attempting to limit or ban these sites, this is seldom successful. The reality is that many people use these sites to gamble on casino like games and many of these can be extremely addictive. Gambling servers are often located in other countries with no restrictive laws making it extremely difficult to control and manage. Young people can be easily lured to the potential of easy money, and the adrenalin rush of winning or losing.

There is a considerable body of work on the psychology of gambling and how to hook gamblers into gambling mire and longer. Online gambling uses much of this theory to make an experience which is irresistible to vulnerable people.

Time Management Apps

Many students, and their lectures have begun to realise that smartphone usage needs to be controlled and managed. A number of attractive mobile apps have recently been developed that allow users to monitor their device usage. For example, an app developed for environmentally friendly students is called Forest. This gives the student a self-administered target during which the student does not access their social media and thus concentrates on their studies. The company gives a score to the user, and plants trees when a threshold is crossed.

Web Browsers

Web browsing on smartphones was originally a painful experience because of the small screens and low computing power of the phone. Many phones now have relatively large and high-resolution displays and much more computing power. The current iPhone X has a resolution of 2436-by-1125-pixels and a 2.39 GHz hexa-core 64-bit processor, in short a very powerful device which can easily display complex web pages with video. Any web page can be displayed quickly, anywhere a student wants to view it.

While there are many excellent educational resources for students, there are many tempting web pages that provide interest and stimulation. Why should a student listen to a teacher when such treasures are available?

What's Popular?

The popularity of smartphone apps can be measured in several ways. First there is the number of downloads reported by the app stores Google Play and the Apple Store. Each app provider also claims a certain number of active or registered users. Without the detailed information available to the app providers or Cambridge Analytica (Granville 2018) it is hard to know if the student body deviates from these global measures of app usage. From our own observations these statistics are about right.

In 2018 the most downloaded apps for Android and iOS were in order: Facebook Messenger, Facebook, WhatsApp Messenger, Tik Tok, and Instagram (Tobin 2018). Of the top 5 downloads five are social media style apps which indicates the main interests of smartphone users.

Summary

There are many social media, entertainment, educational apps, and web sites that are easily accessible to us and our students on a smartphone. To ensure that our students are learning, it is important to understand how and why these apps are used by the students. It is also important to determine the extent of engagement of our students to these sites and apps. A teacher needs to observe what their students are using and try to enter a dialog with students to better understand smartphone usage and student motivations.

The dominant apps used by people of all ages appears to be social media apps which indicates a need for connectivity, interest, and stimulation. Ideally an educational activity can provide these needs and so reduce the temptation to ignore that activity and indulge in the favourite smartphone app.

2.2 What Are the Effects of Smartphones on People?

Students all over the world are spending a lot of time on their smartphones, and some studies suggest that many of us are addicted to these devices. Are smartphones making us smarter, or are we just getting impatient and hooked on constant stimulation? Are we getting better connected and more knowledgeable, or are we getting disconnected from our real local world and environment? Are we now unable to retain information? This chapter examines some of the effects of smartphones on us humans, with the specific emphasis on the students in our classes.

In the early 2000s smartphones started with the Blackberry and the Nokia Symbian based devices which featured not only a traditional cell phone but basic Internet connectivity. In 2007 the first iPhone went on sale. In only a few years smartphones became a "must have" tool that has become the backbone of many people's lives. It supports our everyday activities, such as reading emails, reading books, remembering names and telephone numbers, finding the location of places and even for finding the fastest way to travel to our friend's houses. When we travel, our smartphone does translation, can locate our hotel, and help get transport using apps such as Uber and Ola. Much of the business world has been completely changed by the use of smartphones. The reality is that most of us do not even remember the telephone numbers of the members of our family without the use of the smartphone.

Smartphones have been designed to be convenient to use and a lot of effort has gone into making them pleasingly interactive for a huge range of people. Recently, we met a group of people who were illiterate and unable to sign, but very comfortable in using their smartphones. The apps they used were visual, audio, and icon based and provided this group with much valued applications. Smartphones and their apps break the barriers of language, literacy, are intuitive to use, and have the capacity to adapt to their user. Smartphones and their apps can be used by children, by the elderly, and even by the vision impaired.

The proliferation of these devices has affected every aspect of our lives, and have penetrated into under privileged parts of the world. Cheap basic cell phones are about US$8, basic smartphones around US$40. Recycled phones are common and can be even cheaper. Smartphones have changed human behaviour, the way we do things, and the way we respond. It appears to have changed the way how we seek and respond to questions, and information, and our concepts of knowledge itself seems to be changing very rapidly. As academics, it is essential to understand the changes in our students' expectations and abilities. When looking at the effect of smartphones, it is quite natural to investigate the effect of social media.

Literature

The results of a study (Gladden 2018) have shown that the majority of respondents spend 3–5 h a day using their smartphones, with 75% of respondents saying that they used Instagram and Facebook most. The conclusions of the study indicate that individuals in the 18–30 age range are the ones most affected in their social interactions because of the reliance on smartphones. These younger individuals use smartphones as their main form of communication.

Sarwar and Soomoror (2018), have listed some of the positive effects of smartphones:

1. Improved communication capacity—ability to be in touch with family and friends.
2. Ability to be safe during travel.
3. Ability to drive safely without getting lost.
4. Ability to access healthcare.
5. Ability to measure one's daily exercise.

However, there appear to be many negative effects as well. The problem someindividuals face with smartphones is that they are no longer able to hold face-to-face conversations or communicate effectively in person because they have been communicating solely through the systems available on the smartphones. However, while smartphones are able to provide text, verbal and even video connectivity, the lack of the two individuals being in different physical spaces can lead to incorrect interpretations. Thus, individuals can miscommunicate and further, misunderstand information (Kelly and Miller-Ott 2018). This is extremely prevalent in the cases of younger adults and teenagers who use cell phones more frequently than others and are quite dependent on the devices (Aljomaa et al. 2016). Individuals rely on their smartphones for to communicate through different mediums. As stated by (Katz and Aakhus 2002), the novelty of mobile phone technology, and its intrusive power into people's lives, allow us to observe aspects of the human communication process that would otherwise escape our attention, or at least be extremely difficult to discern. There have been social anxiety and loneliness accusations said to come from too much smartphone and social media use (Gao et al. 2016).

Social Media

Social media has provided the ability to connect with people across the globe in an easy and intuitive manner which was never possible before. Now, families 'meet' on videos while they may be separated by thousands of miles. Classmates can find each other and keep in touch after years of graduation, even when they are located in different countries. We can even measure how many steps our friends took today, the number of calories they consumed and how much exercise they need to do to remain healthy. Parents can monitor the safety of their children while they walk to the school, or in public transport or in taxis. However, it seems that these benefits come with 'side-effects'. The negative impacts of social media platforms include issues like increased levels of anxiety and depression. Research is showing that these can cause poor sleep quality, dissatisfaction with body image, and cyberbullying (Aboujaoude et al. 2015; Lee and Shin 2017; Brewer and Kerslake 2015). A new phenomenon, Fear of Missing Out (FOMO), was of little interest until the availability of smartphones and social media platforms (Beyens et al. 2016).

The reality is that young people spend a lot of time on social media which has become their primary method of connecting with other people. Studies have found the associations between increased social media use and depression, anxiety, sleep problems, eating concerns, and suicide risk. Suzane Fores has made a comprehensive list of the combined negative based effects of smartphones and social media which are listed below:

1. Sleep
2. Romantic
3. Friends
4. Parenting
5. Inability to communicate
6. Expectations of being always available
7. Self-worth measured openly by social media
8. Fear of missing out (FOMO)
9. Reading quality
10. Loss of writing ability
11. Mental health-anxiety
12. Urgency and loss of patience.

These can be grouped in two major categories: Distraction and Emotional. As academics, we have to be aware of these problems, and find methods to mitigate the negative effects.

Distraction

Research has suggested that the social media usage by young adults studying in Universities and colleges has the potential to distract them during their studies, and thus affect their grades (Beland and Murphy 2016; David et al. 2015). It also affects their social-skills and time-management skills. However, it also has positive effects such as makes them have the ability to seek facts and connect with other members of the class. But it appears that in general, many students suffer from their social media usage, though there are a handful that aren't adversely affected and seem to manage social media very well. There are also number of Apps (such as SelfControl) that have been developed that allow the students to monitor and manage their usage and to automatically block the social media during their studies and other activities that are important to them. These are typically designed to prevent the distraction. Selfcontrol and similar apps have anecdotally been found to be very effective in helping students. We are not aware of any university or college which discusses this problem and suggests solutions such as Selfcontrol. It would make a very interesting study to measure the effects of such apps on the student body.

Emotional

The social media is used for a range of things such as keeping in touch with friends and family, finding new friends, flirting, finding life partners, sharing photos and information, for work and for emergencies. It is commonly used to advertise or identify social events and even for selling or shopping goods and services. Thus, in theory, it can prevent loneliness and help users be better connected. However, research is showing that social media can cause anxiety and stress, and in extreme cases has resulted even in death (Vannucci et al. 2017). It is essential to realise the underlying causes of these negative effects and how these could be affecting our students. The management of the emotional impact of social media appears to be more complex than the ones due to distraction, and many of the academics will not be trained and comfortable in handling these issues. Further, most educational institutions are ill equipped to help students handle emotional issues. As academics, we need to get aware of these problems and while this book is unable to give direct strategies, the intent is to help create an awareness and start a dialogue. In our own university there is a competent student counselling group who have been of great value to many students. As an academic with no real expertise in this area our job is to encourage students to go and chat with these people who can offer professional help.

Smartphones can have a big emotional effect on their users

Our experience matches the literature quite well. Students who are motivated and who can handle delayed gratification can benefit from smartphone use both in their social lives and their academic lives. Students who are easily distracted or have trouble with delayed gratification tend to suffer from the use of their smartphones.

Boring?

Most jobs have elements which are repetitive and boring, for example checking calculations, inspecting items, collating and massaging data to make it usable, working through emails, and repeating tests. Meetings can be extremely boring, and the mind can start to wander. In all these cases a high degree of concentration is required to detect errors, complete a task properly, and notice new or unexpected information. In education there are tasks that can be boring; working through practice problems, adding proper references to reports, reading marking guidelines, and much more.

The use of smartphones provides constant stimulation and as discussed can be quite addictive. Given this mental addiction to stimulation, will students or even graduates be willing or able to handle the boring jobs that earlier generations found more palatable? Our own experience is that students are less willing to work through problems, pre-read material before labs and tutorials, or read further to extend their knowledge. The cause is quite obvious, when student attention wavers it is to turn to their smartphone.

The education process must take into account the distraction available with smartphones and we cannot expect younger students to accept boring work in the same way as cohorts from much earlier years. This may prove a problem for industry as our graduates cope less well with boring work than did earlier generations.

References

Aboujaoude, E., Savage, M. W., Starcevic, V., & Salame, W. O. (2015). Cyberbullying: Review of an old problem gone viral. *Journal of Adolescent Health, 57*(1), 10–18.

Aljomaa, S. S., Qudah, M. F. A., Albursan, I. S., Bakhiet, S. F., & Abduljabbar, A. S. (2016). Smartphone addiction among university students in the light of some variables. *Computers in Human Behavior, 61,* 155–164.

Beland, L. P., & Murphy, R. (2016). Ill communication: technology, distraction & student performance. *Labour Economics, 41*, 61–76.

Beyens, I., Frison, E., & Eggermont, S. (2016). "I don't want to miss a thing": Adolescents' fear of missing out and its relationship to adolescents' social needs, Facebook use, and Facebook related stress. *Computers in Human Behavior, 64,* 1–8.

Brewer, G., & Kerslake, J. (2015). Cyberbullying, self-esteem, empathy and loneliness. *Computers in Human Behavior, 48,* 255–260.

Brooks, D. (2017, November 20). Opinion. How evil is tech? *The New York Times*. Retrieved from https://www.nytimes.com/2017/11/20/opinion/how-evil-is-tech.html.

David, P., Kim, J. H., Brickman, J. S., Ran, W., & Curtis, C. M. (2015). Mobile phone distraction while studying. *New Media & Society, 17*(10), 1661–1679.

Farmer, R. (2003, October). Instant messaging–collaborative tool or educator's nightmare. In *The North American Web-based Learning Conference (NAWeb 2003).*

Gao, Y., Li, A., Zhu, T., Liu, X., & Liu, X. (2016). How smartphone usage correlates with social anxiety and loneliness. *PeerJ, 4,* e2197.
Geser, H. (2006). Are girls (even) more addicted. Some gender patterns.
Gladden, D. (2018). The effects of smartphones on social lives: How they affect our social interactions and attitudes. OTS master's level projects & papers. Retrieved from https://digitalcommons.odu.edu/ots_masters_projects/586.
Granville, K. (2018, March 19). Facebook and Cambridge analytica: What you need to know as fallout widens. *The New York Times*. Retrieved from https://www.nytimes.com/2018/03/19/technology/facebook-cambridge-analytica-explained.html.
Katz, J. E., & Aakhus, M. (Eds.). (2002). *Perpetual contact: Mobile communication, private talk, public performance*. Cambridge: Cambridge University Press.
Kelly, L., & Miller-Ott, A. E. (2018). Perceived miscommunication in friends' and romantic partners' texted conversations. *Southern Communication Journal, 83*(4), 267–280.
Kim, H. (2013). Exercise rehabilitation for smartphone addiction. *Journal of Exercise Rehabilitation, 9*(6), 500–505. https://doi.org/10.12965/jer.130080.
Lee, C., & Shin, N. (2017). Prevalence of cyberbullying and predictors of cyberbullying perpetration among Korean adolescents. *Computers in Human Behavior, 68,* 352–358.
Mobile Consumer Survey. (2018). Deloitte Australia. Technology, Media & Telecommunications, Mobile, Trends. (n.d.). Retrieved June 24, 2019, from https://www2.deloitte.com/au/mobile-consumer-survey.
Ng, B. D., & Wiemer-Hastings, P. (2005). Addiction to the internet and online gaming. *Cyberpsychology & Behavior, 8*(2), 110–113.
Parkin, S. (2018, March 4). Has dopamine got us hooked on tech? *The Observer*. Retrieved from https://www.theguardian.com/technology/2018/mar/04/has-dopamine-got-us-hooked-on-tech-facebook-apps-addiction.
Romaniuk, S. N., & Burgers, T. (2019). How the US military is using "violent, chaotic, beautiful" video games to train soldiers. Retrieved July 4, 2019, from The Conversation website: http://theconversation.com/how-the-us-military-is-using-violent-chaotic-beautiful-video-gamesto-train-soldiers-73826.
Sarwar, M. & Soomro, T. (2013). Impact of Smartphone's on Society. *European Journal of Scientific Research.*
Smythe, J. W., & Costall, B. (2003). Mobile phone use facilitates memory in male, but not female, subjects. *NeuroReport, 14*(2), 243–246.
Tobin, A. (2018). These were the most downloaded and profitable games and apps of 2018. Retrieved July 28, 2019, from Forbes website: https://www.forbes.com/sites/annatobin/2018/12/21/these-were-the-most-downloaded-andprofitable-games-and-apps-of-2018/.
Tyler, J. R., & Tang, J. C. (2003). When can i expect an email response? A study of rhythms in email usage. In K. Kuutti, E. H. Karsten, G. Fitzpatrick, P. Dourish, & K. Schmidt (Eds.), *ECSCW 2003* (pp. 239–258). Netherlands: Springer.
Vannucci, A., Flannery, K. M., & Ohannessian, C. M. (2017). Social media use and anxiety in emerging adults. *Journal of Affective Disorders, 207,* 163–166.
Wardle, H., Moody, A., Griffiths, M., Orford, J., & Volberg, R. (2011). Defining the online gambler and patterns of behaviour integration: Evidence from the British Gambling prevalence survey 2010. *International Gambling Studies, 11*(3), 339–356.

Chapter 3
The Effect of Smartphones on Traditional Education

Smartphones have changed student behaviour and expectations, and this continues to evolve. It has also changed the rules of engagement in the classrooms, changed the relationship between students, and between students and the teacher. This section looks at how the smartphones are affecting the traditional classroom and brings to light to the challenges facing both students and academics.

For many generations the classroom structure was static, the academic imparted knowledge to the students with the academic in the front of the class and the students sitting facing the blackboard. With technology, the blackboard changed to become a white board, and then came computers, and tools such as PowerPoint became the standard. While there are differences in the look of classrooms between different Universities, and countries, the fundamentals have rarely changed. The flipped classroom is a new and different approach that is gaining popularity (Gilboy et al. 2015). This requires students to pre-study material (Gross et al. 2015) and then engage with the teacher and other students in the classroom solving problems or discussing the topic studied. Problem Based Learning (PBL) is another new innovation whereby students are given a problem to solve which requires them to master a body of knowledge and possible methods (Savery 2015; Radcliffe and Kumar 2016; Kumar and Radcliffe 2017). The classroom then becomes a forum for exploring what has been discovered, what is still missing, and how to move forward. In all classroom settings with the wide availability of smartphones can cause the dynamics of the classroom to change.

D. K. Kumar and P. Radcliffe, *Teaching Surrounded by Smart Phones*,
SpringerBriefs in Applied Sciences and Technology,
https://doi.org/10.1007/978-981-15-1401-2_3

Traditional Education

A traditional classroom

In today's class, with all the students having access to the Internet while sitting in the classroom, many academics are finding themselves in an unsure position. There are now a number of possibilities that were not there in the past such as the students interacting with other people, be these people within the same class, or outside. Then, they can obtain information from other sources, which may help their learning, but may also challenge the authority of the teacher. And, then there is the problem of managing the students while they need to be doing work such as an exam, or an assignment.

Student use of smartphones changes the effectiveness and outcome of traditional educational methods. The students can be distracted because they may be engaged with the social media, or they simply would find the academic topic dry, or difficult and drift into their favourite app, game or web site. Then, some of the smart students may attempt to go online to check the topic from other sources, which, while may sound useful, could cause disruptions to the classroom. But perhaps the biggest challenge is the change in the expectations of the students. Thanks to the ever-present smartphone, students expect to have the information now, without waiting and without patience. We do know that information can be provided without delay, but to develop the expertise, to make a person an expert on the concepts underlying the information requires significantly more time and effort.

To evolve a person to think independently is even more difficult, both, for the teacher, and for also the student. Smartphones can provide quick answers to nearly anything, but this does not make them independent. Often, this may give the appearance of independence because the students do not seem to need the information provided by the teacher, as they can access the entire global data. However, it may be just the opposite, because the students get very dependent on the information that is available using their smartphones, they can lack the ability to reason and question. Such students would not know how to evaluate information critically, investigate for a range of solutions, and perform problem solving, that is outside what they are able to find online.

Thus, smartphones are making it very challenging for the traditional teaching model in the classrooms. When the students are using these smartphones without proper directions, the learning outcomes can be significantly poorer. While they are able to get quick information, it makes it difficult for the teacher to challenge them to develop the necessary skills for analysis, problem solving or the understanding of abstract concepts.

Distraction

The issue of distraction has been mentioned, but is any use of a smartphone distracting or only some types of app? Work from David et al. (2015) suggests that in the home study situation, social media apps are a significant problem but listening to music does little to interfere with study. Smartphone use is definitely linked to poorer performance in the classroom (Junco 2012; Rosen et al. 2013; Wood et al. 2012). Heavy users of smartphones who heavily multitask have been found to have problems in focusing on their work (Ophir et al. 2009). Multitasking whilst studying is linked with shallower processing (Carr 2011).

Our own experience is much along the same lines that students who heavily use their mobile phones get little value from an educational activity. Particularly first year university students have just come from high school where attendance is compulsory, for the body perhaps but not necessarily the mind. This same attitude is brought to university, smartphones now enter the equation, and so many students feel they have attended and satisfied some kind of requirement just by being physically present. Such heavy smartphone users may as well have stayed away as they do not engage and seldom retain much in the way of knowledge or methods.

Smartphones can be very distracting

Some students do not suffer from the use of smartphones. These tend to be the more capable students who carefully choose *when* to use their smartphone. These students tend to be easy to identify as smartphone use stops when key material is being delivered or students are required to work on a problem. At least in the STEM area we have found student multitasking between smartphone and an educational activity does not work well except perhaps in lecturers when the student can easily follow the material.

Lectures

Long before smartphones were available students had to copy down lecturers in their own handwriting on paper. This kept students very busy and attentive and helped to cement the material into memory. A failure to copy down material would often mean that labs and tuts became impossible and the exams would be a disaster. The most extreme case of which we are aware were medical lectures in embryology, it took three students to take notes, one for what the lecturer wrote, one for what they said, and a third to copy down the diagrams.

Many universities then started handing out notes and recording the lectures for students to review later using their desktop, laptop computers, and more recently smartphone. This brought many advantages to students, but lecturers were left with a less attentive and often much smaller audience. Some students, noticeably those with lower motivation, did not attend lectures intending to view the video later, but never got around to it. Our experience is that these students were often totally unprepared for the following labs and tutorials.

None of this behaviour is the fault of smartphones and so they cannot be blamed for this problem.

There are several smartphone problems which do affect lectures. The first is the issue of multitasking. Humans are quite capable of multitasking between different tasks but the cognitive power available to each task is severely lowered compared to giving all one's attention to a single task (Dux et al. 2009; Koch et al. 2018). Dux et al. also claim that training can improve the multitasking capability. In a typical lecture there are times when there is little happening, for example waiting for the lecture to start, and using a smartphone has a distinct advantage. As a lecture progresses the cognitive load increases and the student understanding and retention starts to falter. If the smartphone activity is too attractive, then the lecture is ignored, and its educational value is lost.

Exams and tests

A big challenge for the teacher in the traditional classroom with students having access to their smartphones is for exams or tests. It is extremely difficult for any teacher to effectively regulate and prevent their students to have access to their smartphones in a lab or tutorial, when working on assignments, or even in a test or examination. With the steady technical development, already the size of these devices is now wearable device as a watch or a piece of jewellery. The Apple Smart Watch can be configured to look like an innocent digital watch but in fact can connect to the Internet and have a whole variety of apps running on it. It is effectively possible to "bring a buddy" into the exam with you by having a friend on the other end of a text chat app.

Looks just like a watch! Actually, it's an Internet connected screen and input device!

For the purposes of exams most universities have responded by publicizing new rules and sanctions, and then applying those sanctions when perpetrators are caught. The publicity is relatively easy with students being clearly informed that smartphones and watches are banned in tests and exams. The weakness is in the ability to detect such devices. Exam invigilators need to be trained and tutors may need training as well. To stop wholesale cheating students must see there is a high chance of being caught and that there will be unpalatable consequences when caught.

Take home assignments

Take home assignments have been controversial for some time as they are commonly group worked and cheating is very easy. Some 40% of university students self-report cheating (Plagiarism: Facts & Stats 2019), given this is self-reporting the real rates are likely to be much higher. If you are in doubt of these figures do a web search with the terms: do my assignment for me. You will see a huge number of enterprises who will do your assignment for you, often at trivial rates. Some even advertise "HD or your money back". The number of enterprises alone speaks of a flourishing business.

Any mark from a take home assignment is of very dubious value unless there are is some attempt to individualize the assignment. The problem is particularly severe when there is one answer which is the same for all students. In one of our courses on digital design the problem given has around 60 variations depending on the student number and their age. We have shown that this improved the correlation between lab marks and exam marks, which strongly suggests there was less cheating, students had to do the work themselves, and so did better on the exam.

What role does the smartphone have in take home assignments? It acts as an enabler for getting easy and rapid help and so diminishes the threshold at which students will give up and try to get the answer from another student. Without a smartphone a student may have to wait until the next day to get help and so is more likely to try a little more themselves but when help is easily on tap why bother?

The flipped classroom

The flipped classroom approach requires students to prepare before the class activity and then discuss the educational material in the class activity (Gross et al. 2015). Any lecture becomes more of a lectorial which is interactive, solves student problems, and often extends on the material students have already reviewed (Bishop and Verleger 2013). This comes at the cost of greater staff labour in preparation, student acceptance of a new paradigm including preparation before a class activity, and concerns about covering enough material Herreid and Schiller (2013).

Smartphones can both help and hinder this activity. Students may not have completed the required preparation before the class. With a smartphone they can at least start preparation on the way to class or before the class. We have observed on public transport that many people are working on their mobile phones whilst on their way to or from work and students are no exception. Smartphones in the classroom can be more of a problem. Students who have not prepared tend to keep on preparing and do not join in the discussion. The immense distraction provided by a smartphone has already been discussed and contributes to many students failing to engage in the classroom. This issue is discussed further in one of the case studies presented in a later chapter.

Styles of learning

It has long been claimed that students will have different preferred forms of learning such as visual, textural, kinaesthetic, and auditory. It is further suggested that educators should deliver in multiple modes (Reid 1995; Schmeck 2013). While this may enhance the education process it may not get the student ready for the workplace and so consideration should be given to teaching in the modes commonly found in the workplace.

The smartphone can be of great help here as it can convey resources that are visual and auditory in nature. There may well be significant material already available which may reduce the workload on educational staff. There is still a significant workload in finding the material and regularly checking it is still available. Producing new material of this nature can be very time consuming and requires planning and time allocations for staff.

This section has already alluded to the strong distracting power of smartphones and so how this material is delivered must be carefully planned. Some resources might be accessed by students outside of the classroom. If the material is delivered in the classroom then the approach of "start your phones" and then firmly "put down your phones" may be effective if groups that are smaller and where the teacher can move around the class.

3.1 Access and Equity

This section looks at the question of equity and access. What is the difference between $2000 smartphones and cheaper models? Might that affect the educational outcome? How should the teacher plan to ensure that this difference will not affect the learning outcomes of the students? Can all students use smartphones, or will some have difficulties?

Affordable?

Until recently it could be claimed that there would be some students in difficult financial positions who could not afford a smartphone. In the current climate this assertion is most likely false. Our experience in a number of countries is that smartphones are so essential to young people that everyone has one. According to an analysis by Forbes of various reports, 38% of Americans would give up sex to keep their mobile phones (Mills 2019). Given the overwhelming urge to have a smartphone nearly all young people have found a way to obtain one.

We know of one good student in dire financial straits. He was given an old smartphone by a friend, but he could not afford a phone plan. He was able to use the phone on the free WIFI available at the university, several shops, and his accommodation. He used apps such as Skype and WhatsApp for social communication and so satisfied his social and academic needs.

A school or university can play a role in ensuring all students have access to some form of mobile phones. One of the best ways is to engage in an active phone recycling system that makes phones freely available to disadvantaged students.

Currently it is safe to conclude that all students have access to a smartphone, but the capabilities of those phones can be dramatically different. Does this difference represent an equity and access issue in a school or university environment?

Technical capabilities

Smartphones are devices that all have an active screen which displays information and receives input from the user. They come with cameras, microphones and speakers, have connectivity to the phone system, and WIFI connectivity. The price range of smartphones is huge ranging from US50 to US2000. In part this price difference can be traced to brand name and status but there are very real hardware differences. The first significant difference is the screen in terms of physical size and the number of pixels that make up the screen. Low end phones will have trouble displaying images which are detailed or large. Some smaller phones have real trouble displaying an image and the touch keyboard required to enter text. Processor power is another major difference with the more expensive machines have significantly more processing speed and multiple processor cores. This processing power is especially important when playing any form of video as slower phones cannot display high resolution videos, will drop frames, and may distort the audio. Processing power can also affect the speed of the data link with low end phones having trouble downloading and displaying videos at the same time. In general phone memory is not a problem except in the very lowest phone models which cannot store many applications or large files such as video files. The ability to take good quality pictures or videos is important in some courses. Again, low end phones have markedly worse camera resolution and may struggle when taking videos longer than a few minutes.

The preceding discussion identified the differences between low-end and high-end smartphones and can be used to create guidelines about how educational material should be delivered by smartphones. The first guideline is that any material should display well on small screens. Videos should be available in low resolution formats so they will display well on less capable phones. Activities that require the student to take camera photos or video should accept low quality, and low-resolution images.

There is a lot of guidance on the web on how to make web sites and other resources suitable for low en smartphones. A good start is the Google guide to making web sites mobile friendly (Search Engine Optimization… 2019). It is easier again to use any experts available in your educational institution. Many universities have helpful web content experts who can save you a great deal of time and effort and produce a much better result than you could by yourself. Some universities only allow you to put up material through such experts to ensure a high quality and consistent look and feel to all web-based material. Educational packages such as Canvas automatically handle serving to screens of different sizes.

The ability to use smartphones

No discussion of Access and Equity would be complete without consideration of students who are disabled in some way. Smartphone usage is dependent on the ability to hold the unit, use its touch screen, and see the possibly small display. Various muscular disorders can make operation difficult, and vision disorders may make it difficult or impossible to see a small screen. These problems are often solved at a cost, large tablets make the touch screen much large and so make input easier. The larger screen makes it easier to see any images.

Most smartphones and tablets have a variety of accessibility features such as text to speech and speech input. Digital voice assistants such as Siri and OK Google can achieve a great deal just by voice commands. There are a few simple things that can be done such as installing a magnifier app to make text larger, and simply setting the text to be larger. Such features may result in smartphone resources being easier to use for disabled people than traditional paper and book resources (Rebelo and Soares, 2014).

Most universities have an access and equity group who can give good advice on the current accessibility features and problems with smartphones, tablets, and computers. There is good advice both in the academic and popular literature (Irvine et al. 2014; Android accessibility overview… 2019).

Conclusion

In the current era we can assume that all students can get access to a smartphone which is usable for education. The education resources must be designed to be usable by low end smartphones which have limited size displays and poor computing power. Accessibility of educational material can be improved by using smartphones due to features such as text to speech. Access and equity to educational material can be improved if simple guidelines are followed.

References

Android accessibility overview—Android Accessibility Help. (n.d.). Retrieved July 10, 2019, from https://support.google.com/accessibility/android/answer/6006564?hl=en.

Bishop, J. L., & Verleger, M. A. (2013, June). The flipped classroom: A survey of the research. In *ASEE National Conference Proceedings, Atlanta, GA* (Vol. 30, No. 9, pp. 1–18).

Carr, N. (2011). The shallows: What the internet is doing to our brains. WW Norton & Company.

David, P., Kim, J. H., Brickman, J. S., Ran, W., & Curtis, C. M. (2015). Mobile phone distraction while studying. *New Media & Society, 17*(10), 1661–1679.

Dux, P. E., Tombu, M. N., Harrison, S., Rogers, B. P., Tong, F., & Marois, R. (2009). Training improves multitasking performance by increasing the speed of information processing in human prefrontal cortex. *Neuron, 63*(1), 127–138.

Gilboy, M. B., Heinerichs, S., & Pazzaglia, G. (2015). Enhancing student engagement using the flipped classroom. *Journal of Nutrition Education and Behavior, 47*(1), 109–114.

Gross, D., Pietri, E. S., Anderson, G., Moyano-Camihort, K., & Graham, M. J. (2015). Increased preclass preparation underlies student outcome improvement in the flipped classroom. *CBE—Life Sciences Education, 14*(4), ar36.

Herreid, C. F., & Schiller, N. A. (2013). Case studies and the flipped classroom. *Journal of College Science Teaching, 42*(5), 62–66.

Irvine, D., Zemke, A., Pusateri, G., Gerlach, L., Chun, R., & Jay, W. M. (2014). Tablet and smartphone accessibility features in the low vision rehabilitation. *Neuro-Ophthalmology, 38*(2), 53–59. https://doi.org/10.3109/01658107.2013.874448.

Junco, R. (2012). In-class multitasking and academic performance. *Computers in Human Behavior, 28*(6), 2236–2243. https://doi.org/10.1016/j.chb.2012.06.031.

Koch, I., Poljac, E., Müller, H., & Kiesel, A. (2018). Cognitive structure, flexibility, and plasticity in human multitasking—An integrative review of dual-task and task-switching research. *Psychological Bulletin, 144*(6), 557.

Kumar, D., & Radcliffe, P. (2017). Problem based learning for engineering. *In 2017 39th Annual International Conference of the IEEE Engineering in Medicine and Biology Society (EMBC), IEEE* (pp. 25–29).

Mills, M. P. (2019). *The mobile revolution has only just begun*. Retrieved July 10, 2019, from Forbes website: https://www.forbes.com/sites/markpmills/2015/01/19/the-mobile-revolution-has-only-just-begun/.

Ophir, E., Nass, C., & Wagner, A. D. (2009). Cognitive control in media multitaskers. *Proceedings of the National Academy of Sciences, 106*(37), 15583–15587.

Plagiarism: Facts & Stats—Plagiarism.org. (n.d.). Retrieved July 13, 2019, from http://www.plagiarism.org/article/plagiarism-facts-and-stats.

Rebelo, F., & Soares, M. (2014). Advances in ergonomics in design, usability & special populations: Part II. AHFE Conference.

Radcliffe, P. J., & Kumar, D. (2016). Is problem-based learning suitable for engineering?. *Australasian Journal of Engineering Education, 21*(2), 81–88.

Reid, J. M. (1995). *Learning styles in the ESL/EFL classroom*. Heinle & Heinle Publishers, International Thomson Publishing Book Distribution Center, 7625 Empire Drive, Florence, KY 41042.

Rosen, L. D., Mark Carrier, L., & Cheever, N. A. (2013). Facebook and texting made me do it: Media-induced task-switching while studying. *Computers in Human Behavior, 29*(3), 948–958. https://doi.org/10.1016/j.chb.2012.12.001.

Savery, J. R. (2015). Overview of problem-based learning: Definitions and distinctions. *Essential Readings in Problem-Based Learning: Exploring and Extending the Legacy of Howard S. Barrows, 9,* 5–15.

Schmeck, R. R. (Ed.). (2013). *Learning strategies and learning styles*. Springer Science & Business Media.

Search Engine Optimization. (SEO). *Starter guide—Search console help*. (n.d.). Retrieved July 10, 2019, from https://support.google.com/webmasters/answer/7451184?hl=en.

Wood, E., Zivcakova, L., Gentile, P., Archer, K., De Pasquale, D., & Nosko, A. (2012). Examining the impact of off-task multi-tasking with technology on real-time classroom learning. *Computers & Education, 58*(1), 365–374.

Chapter 4
Using Smartphone to Enhance Education

With the growth of smartphones and associated technologies has come millions of smartphone programs which are called "apps". Most apps are like any tool, if used sensibly they can be of value, but if used unwisely they can distract students to the detriment of their education. Many techniques that use smartphones have been proposed to help improve student learning and for better educational outcomes. This chapter looks at some of these methods that can help nullify the negative effects of smartphones in the classroom or co-opt them into helping the education process.

The dangers of projection

One of the biggest dangers when selecting tools, methods, and techniques for education is projection where the educator projects their interests and values onto the student body (Kumar and Radcliffe 2019). This can result in predictions of student behaviour that are wildly different to what will actually occur and consequentially a disastrous education activity. We have seen educators who believe very few students will cheat, a view roundly rejected by real measurement (Plagiarism: Facts & Stats 2019) with typically 60% of high school students and 40% of college student self-reporting as cheating. A very common delusion is that students, like the educator, will see material as being interesting and useful where in fact they see it as boring and only necessary to pass a test or exam. Before selecting smartphone tools and methods, it is essential to first observe the student body in question and interact with that body. The views of tutors can be very valuable as students can behave quite differently when academics and teachers are around but are usually more honest with their tutors who are often closer to their own age. Many groups have a variety of student types. We have found at least three types of students in most of our larger lectures, with the proportions varying considerably depending on age and past performance. We are dubious that educational experts who do not have a current, active teaching role can accurately model students. Experience and personal contact are required to make these judgements.

D. K. Kumar and P. Radcliffe, *Teaching Surrounded by Smart Phones*,
SpringerBriefs in Applied Sciences and Technology,
https://doi.org/10.1007/978-981-15-1401-2_4

Projection in education: what I value, the student values

Once the educator has a mental model of the student body including its variations, then the educator is in a good position to judge if a tool or method will work. For example, in one first year group we were considering whether to use smartphones in tutorials. We went through the process of characterizing the student body and concluded that this would be a disaster. Why? The course was a service course which had little to do with the main thrust and interest of the students and their degree program. The students were all 17 or 18 years old fresh out of high school and eminently distractible. It was felt that if smartphones came out for legitimate reasons they would soon be used for non-educational reasons and distract the students from the educational activities.

4.1 Some Useful Tools

As of the first quarter of 2019 there are approximately 2 million apps each on Google Play and the Apple App store (App Stores 2019). Some apps are very useful for students, academics, and administrators. Each app has their own purpose and target, and many of these are free. For the administration domain these can be broadly categorised in four categories:

1. Administration and management.
2. Active classroom engagement.
3. Alternate contact method.
4. Providing students with learning material. Most tools include this feature.

Most of these apps have been designed for non-technical people who should be able to use the app with little or no technical help. In general, the setup and use of these tools is not that difficult except for the creation of educational material which can take a great deal of time, far more time than the creation of traditional educational material especially if marking or evaluation is to be automated as well. The educator must ensure they have sufficient time allowance for this task or face a huge time impost on their personal lives. It can be extremely useful to chat with experienced users to understand their problems, the effort it takes to use such tools, and the real benefits to the education process.

Most educational contexts have a target output or employment environment for the students. In a university context, the output environment is usually a professional work environment which will have its own set of work tools, expectations, and methods. In order to make students more ready for life and work the educational methods should have some match to this output environment otherwise graduates will not be perceived as work ready by employers and have trouble getting a job. Students can find it very motivating if they understand the education methods being used will help their employment. The application of smartphones in education should be attuned to work in harmony with this goal and should match workplace use.

The pain of early adopters can be a useful way to evaluate any tool. The web is filled with reviews of various tools, but it can be very difficult to determine if these are honest reports or whether the reporter has a vested or financial interest which has biased their opinion. Sadly, buying reviews and opinions has become common place and the reader needs to be ever more vigilant to try to find an honest opinion. In general, we have found the following styles of reviews to be biased and of limited use

- Anything released by a company who owns or distributes the product, or a competitor of a product.
- Overly positive reviews that do not mention any negative features.
- Reviews that neglect free or open source products.
- Reviews that do not clearly come from an end user.

Administration and management

Administration and management apps aim to help the academic manage students effectively. In this group are apps that are suitable for managing student attendance. There are also apps that can be used by students to submit work, and the academics to provide students with the marks and individual and group feedback. Some tools also test if there was any copying and the level of similarity between the submission of different students, or with some material available online, any of which could identify plagiarism. Some of the well-known apps and software are:

Google Classroom

G Suite for Education (formerly known as Google Apps for Education) has a range of tools, one of which is Classroom. This allows the teacher to receive individual assignments, give individual feedback, and distribute grades through the app. It also facilitates sharing class materials using platform such as Google Drive, and make announcements. Another option for the use of this is to open common documents and thus engage them in discussions.

Canvas

Canvas is a complete Learning Management System (LMS) which provides a complete web-based solution for the management and delivery of educational material, marking and feedback. It is a complex package that requires an expert team to support and manage the servers and guide academics and teachers on how to use the system. It can be used to make announcements, share course material, for student submission of assignments, and to give students feedback and grades, while maintaining the confidentiality of each student. Canvas appears to be the most popular LMS in the western world.

Blackboard

Blackboard is much like Canvas, a huge and complete LMS. Again, it requires a team to support the product.

Moodle

Moodle is yet another LMS but this one is free and open source. I can be run on any platform capable of running an Apache web server. Some academics have run the basic system on their own server, but it is much better to have a support team. There is excellent support available from several commercial groups as well as forums.

Additio

This software allows an academic to keep attendance, grades, and class notes. It contains a digital gradebook and classroom management app which facilitates recording attendance and help students to manage their timetable on their smartphone.

Turnitin

Turnitin is a well-known package that does comprehensive checking of a student's assignment against other submitted assignments and the whole web. The tool can handle assignment submission, plagiarism detection and reporting to the academic and the student. The entire system is easy to set up by a teacher or academic and does not require a support team.

Active classroom engagement

This type of software and apps can be used by academics, teachers, and students to help the students have a better learning outcome. There is a wide variety of these apps including tools to make the students more engaged in the class or provide learning material or leads to the learning material. There are also apps that allow the students to get their smartphones or laptops connected with the classroom screens and handwrite on the screen using their smartphones. Examples of the commonly used tools are:

Kahoot

Designed to help teachers motivate their students and engage them in the class. The website, GetKahoot offers some screen designs using which the academic can enter questions and material into the site to create an interactive 'game' like situation. It provides students with the Kahoot app which can be used as a buzzer and for voting for a response. Student scores, opinions, and text replies can be seen on the classroom screen.

In our experience, free text replies should be avoided. In a larger class there is always someone who will abuse the anonymity and post inappropriate comments.

One-Drive

Similar to the Google platform, this is the platform from Microsoft, which allows the academic and the students to share files. It allows multiple users to open the files simultaneously and can be useful for teamwork. The academic can select people who have access to the files, and it provides a measure of the activity from individuals.

EduCreation

This is an interactive whiteboard app which provides students and teachers the ability to record audio, create animations, write comments, draw diagrams and instructions.

Alternate contact method

There are a large number of apps that allow Schools and Universities to contact their students, and the families, if appropriate. These may be to make announcements, provide grades, or even deliver educational material.

SeeSaw

This is designed for Schools rather than Universities, and it allows the parents to monitor the progress of their children. This helps create a portfolio of the child which can be observed by the parents and is designed to reduce the need for the face-to-face meetings between the parents and the teachers.

Remind

This is designed for teachers to communicate with students. It allows the formation of groups, and announcements can be made to the group. It can also be used to initiate group discussions. One big advantage of this platform is that it allows automatic translation and thus is suitable when the class may have large number of people who are non-native English speakers.

Slack

This is designed to help academics contact students for hosting discussion forums and pushing important reminders. It also allows academics to monitor the group discussions and follow the discussion threads.

Schoology

This cloud-based platform is to promote communication and networking between classmates and peers while providing security. Thus, while it is similar to other networking forums such as Facebook and WhatsApp, but it claims to provide the security to ensure that the activity can be monitored by the academics and teachers. It provides a platform for students to work in teams while working on projects and assignments. It can also be used for managing assignment submission.

Mindmeister

This app makes it easy to map out the relationships between ideas. For complex group projects, it can also provide a way to quickly visualize and create a project outline, together.

Other Tools

Grammarly

Grammarly is a well-known tool that helps students identify weaknesses in their grammar and suggest how to improve that grammar. It can be used on a range of devices and via the web. Recently Grammarly has added plagiarism detection which can detect copying from almost anywhere on the web. It can also detect where a document has not cited work and provide the citation. This tool is designed to be used by an individual and not to evaluate the submission from an entire class group.

4.2 Tools for Policing the Use of Smartphones

What to do when students have smartphones in the classroom? How can we police the class so that we can reduce the distractions due to these in the class? Recently we asked these questions on two well-known online forums: LinkedIn and Research Gate. The response was very interesting, and we were surprised by the energy and speed of the responses. There were hundreds of people who responded immediately and were suggesting the use of electromagnetic solutions such as use of Faraday cage or radio jamming. Then there were more old-fashioned methods such as the use of strip-search, or threat to fail. I then asked these people if they had implemented these, and the response that they wanted to, but could not.

The reality is that none of them are truly implementable in the modern era. With technical growth, we already have Apple or Huawei smartwatches and embedded smartphones are not too far away. What do we think we can do to police the use of these devices?

The reality is that there are going to be situations when it is highly desirable to stop the students to have access to their smartphones. For example, it will be difficult to ensure exams are fair when students have easy access to their smartphones or similar devices. But the reality is that even when airlines have been threatening the flying public with potential air-crash, many passengers do not bother to turn off their phones.

Policing smartphones can be a lot of effort

Factors to Consider

The solution of using smartphones, banning smartphones, or policing them in the classroom is dependent on a number of factors such as the age and type of students, the course, and size of the classrooms. What may be possible in one situation will be impossible in another. What may be necessary for one cohort may not be required for the other. Again, there is no one answer, but what is possible is to identify some of the factors that may help make the decision. What is the age of the students? What is the relationship of the students to the teacher? But most importantly, how large is the classroom and what is the teacher to student ratio. If all the students in the class are in the direct sight of the lecturer, and there is a high level of respect for the academic by the students in the classroom, then it is likely that most students at most times will not carry their phones into the class. However, a large class in an average University often does not offer this luxury to the academics. It is thus essential to accept the inevitability of smartphones in the classroom and find methods to deal with the problem.

Set the Culture

As with any rules, it is vital to get student acceptance and conformance to rules about smartphones. Most students will respond to arguments which appeal to their "enlightened self-interest". If students can see that the rule helps them avoid pain or gives them an advantage, then they will generally accept the rule. If such arguments are not clearly presented, then the game becomes how to work around the rules. It can also be helpful that any pain is not caused by the teacher but some other party. For example, here is a quite from one of my favourite tutors-" The exam is pretty tough people and I know you will be required to solve problems just like these. So now it's phones off, laptops away, and let's start working through the problems.". The tutor is not seen as the "bad guy", in fact there is no "bad guy", just a danger to overcome. It is important to repeat such messages to ensure students hear and accept those messages. Additionally, everyone in the educational team must articulate and support the message.

The educational environment can be ruined by badly though out rules. For example, rules which are unenforceable or flouted with no consequences, will cause students to lose respect for the teachers and subject. Some rules may be technically enforceable but the workload to enforce those rules becomes unsustainable. In these circumstances it may be better to instead offer strong guidelines or recommendations but leave it up to the student to self-enforce. Again, the use of "enlightened self-interest" can aid student adherence to the recommendations.

Smartphone monitoring

There are a variety of tools that enable the user to monitor their smartphone usage both by application and time. Both Android and iPhone app stores have many phone usage monitors and app monitors. All can display phone usage and some also allow apps to be blocked after a certain period of use or at particular times of day.

There are many apps for young people with parental style controls, but these tend to be a yearly subscription as they rely on black and white lists for apps and websites which must be continually updated. Apps such as App Usage and Quality Time run on the smartphone and are more suited to adult usage.

In a university setting it may only be possible to strongly suggest students use these tools to understand their real phone usage and enhance their academic performance. In a school environment it may be possible for parents and teachers to review phone usage with a view to helping students better control their smartphone use.

Jamming and Blocking

It is possible to disable all smartphones within a given radius by use of a jamming device that interferes with the radio signal. The jamming of mobile phones is illegal in most countries and does have its dangers as no one in the affected area can make calls, even emergency calls. Jamming devices can be quite cheap, in the order of US50, but it is illegal to possess such devices in many countries. Frustration has caused teachers to jam smartphones (Matyszczyk 2019) but this usually has severe negative consequences for the teacher.

Another more sophisticated and expensive approach is to mimic a phone cell tower and only allow certain calls or data through. Devices called Stingrays (Cell-Site Simulators/IMSI Catchers 2017) are already used by police in several countries. Again, such devices are illegal for citizens to use in most countries.

Controlled access: it is possible to require students to logon to an institutions WIFI which can then limit where the students can go and what they can see. This requires the services of an IT section or network engineer who can set up a firewall to achieve this goal. Unless the network access is properly setup, students can switch to accessing the Internet via their mobile phone accounts and then also access anything they want.

This approach may become practical when students need to use a resource which is only available via the institutions intranet and is not available via the public Internet.

4.3 Methods and Techniques

This section looks at some generic ways in which smartphones can be used in education. Previous sections have discussed the many dangers and problems associated with smartphones, and the issue of properly modelling student behaviour and not mistakenly projecting the educator's views and behaviour onto the student body. All the issues must be kept in mind when considering any of the method outlined below.

Some of the methods suggested can be used together, for example reporting via smartphone tools and time limited access.

Ban all smartphones

A ban on smartphone use can only work when the teacher has considerable authority and has the physical freedom to walk around the student group and check on any smartphone use. The authority may come from personal presence or the threat of some type of sanction which matters to the student body. This approach can be aided by formal and well publicized institutional policies for example the government ban of smartphones in primary or secondary schools has already mentioned (Beland and Murphy 2016).

A better outcome is when the students do not feel the need to use a smartphone. The educational activity keeps them busy, or interested, or they are motivated by other reasons such as a coming assignment submission, test, or exam. It is not always possible to motivate such students without excessive preparation workload and/or reducing the rigour of the education process ("dumbing it down").

The ban all approach is of most use when students are easily distracted (most younger people), unmotivated, easily bored, have poor attention span, or seek immediate gratification.

No smartphones, just phones

France has banned smartphones in primary schools and requires manufacturers to offer phone plus text style phones with no advertising (Ledsom 2019). This may be viable in a primary or high school situation, but it is not viable in most universities.

Free access

Free access to smartphones is allowed either by permission of the teacher or because there is not practical way to ban their use. Even though this may not be desirable. In general, it is better not to have a rule, than to have a rule which is flagrantly ignored.

From our experience this strategy is only successful if the student body is highly motivated and interested in the material. This may come from a student's internal drive but may also be driven by external pressures such as a coming exam or test. Smartphone use can be positive if the students are made aware of important resources which they can access. This can be a considerable burden on the teaching staff as they must find existing resources, check old resources are still available and suitable, generate new resources, and maintain resources generated by them.

Time limited access

Students are allowed to use mobile phones for a specific task and must then put them when the teacher requests it. This only tends to be successful in small groups where the educators have some authority. The tasks must be well planned so students have a clear objective and a reasonable time to complete the task. Too much time allows them to stray, too short a time and they get frustrated by not being able to finish the task.

In larger groups, where the teacher cannot move about the room, we have found students keep out their smartphones and use them for other purposes.

Access specific information or tools

Students can be asked to access specific web sites or tools. In small groups this allows the teacher to do some checking if students are being distracted by non-educational activities. The educator must create or find these tools and sites, a process which can take a great deal of effort.

This technique works best if the teacher has some authority and can redirect students who have strayed into non-educational zones.

Report via tool

In this approach a student will report their results using tools available through their smartphone. Many educational management tools also allow the teacher to review and mark the material via the same tool. Such systems require careful set up before the teaching activity but may reduce the effort to mark and return to students. If any form of automated marking is possible then there can be a significant reduction in staff effort (Radcliffe and Rudolph 2007). Other tools such as Turnitin for plagiarism and Grammarly for grammar can also improve the learning activity.

Tools such as Google Forms while not intended as educational delivery tools can also be useful for gaining student input. The powerful spreadsheet functions behind Google Forms can reduce teacher effort.

Self monitoring

In the previous subsection on tools it was noted that there are many tools by which smartphone user can monitor and even control access to apps such as Facebook. At primary school level it may be possible have parental controls but in later high school and university this is not practical. However, teachers and parents can recommend and discuss these apps and show how they work. This can be followed up by asking students how the monitoring and blocking is working.

At minimum students will be more aware of smartphone problems and get a realistic measure of their own use rather than relying on wishful thinking that they are OK.

Do an activity

Smartphone use is more likely to continue in class if the student is only required to listen. If an activity is required, then the majority of students will put down their smartphones and re-join the class. Activities we have used in the lectures and found a success with STEM students include

- Calculations including a challenge to try mental calculation.
- Pose a problem and ask students to tell their neighbours of their opinion/prediction/result.
- Use the smartphone to do a task, followed by a firm instruction to put the phones away.

Group work

Some activities can be turned into group work. For example, groups may be asked to find information together and then write into a report or blog which is later assessed. With careful design an oversight group work can be enjoyable for students.

This approach has some benefits but also some severe limitations. Some students will not work well in a team due to timidity or cultural issues and so their education outcomes may be negative. Dominant students may sideline other students and again ruin the educational outcome. The most important problem is that any group mark very seldom represents the true contribution of each student in the group. In our experience and that of our close peers, there is no reliable way to moderate these marks per student unless there is staff insight into individual student performance. We have found that attempts at peer review without this feature usually fail dismally and may lead to significant injustices. We have also seen significant student anger and disillusionment because some students who have done little work, share a good mark because of the work of others. There is one responsibility all educational institutions share; all students marks should represent that student's skills. Group marks usually fail that test and so should be used sparingly and have low weight.

Discuss when to copy and get help

Smartphones make it very easy to get help and answers. Students live busy lives with many other calls upon their time and other desires to fulfil. In the days before smartphones a student might have to wait a day or more to contact other students and so they were more likely to have a try at the problem themselves. With the massive and instant interconnectivity offered by smartphones and their apps, help is just a few clicks away so why not get that help and so save time and get a better mark?

If there is a good reason to not copy, or to persevere and solve a problem, then the teacher must be able to explain this reason in a way that makes sense from the perspective of student self-interest.

For example, in one of our digital design courses we made it very clear that the final exam would test student's ability to design digital circuits much like those designed in the labs. We reminded the students several times of this fact and encouraged them to master the design method by themselves and not to cheat, because if they cheated then they would not gain those skills and fail the exam. When we introduced this motivation, it appeared to work well, and even weaker students were more diligently in trying to master the techniques. Now a counter example, imagine you are taking a course in a university while you are busy working at the same time, and that course has no relevance to your career or aspirations. The lecturer has set a long and boring assignment which appears to have little learning of any value. The answers have appeared on a Facebook page set up by several other students. Why should you not take advantage of this page? To make it worse everyone else appears to be using it. Would you really not consider using that material?

This discussion goes to the heart of curriculum design in that any educational activity must make sense from the student perspective and that "sense" must be clearly articulated by the teacher.

A key problem with this approach is for the student to know when to give up and ask for help, and what strategies to use. Such knowledge comes naturally to those with years of experience but not to many young people. Our solution was to give students guidelines on when to ask for help

- If the problem looks very big find a bit that you do understand and work on that.
- If you cannot find a place to start review the lecture notes, tutorial work, worked problems, and the references.
- If you are still stuck write down your key problems and ensure they are clearly stated.
- Approach friends and the tutors to try and break through your key problems.
- Use the steps above repeatedly until you arrive at a solution.
- If you have made an honest attempt at a problem, then feel free to check that with friends and the tutor.

Though we have not made any formal assessment of this strategy we have noted students are much more comfortable knowing what is allowed help and what is considered cheating.

Systematic Design of Rules

There is a large body of knowledge about rules and human behaviour when rules are applied. The research done under the social rule system theory (Flam et al. 2008) presents an interesting view of the role of rules in human society. The adherence to rules varies a great deal depending on the cultural background of the students, the rules themselves, the chance and consequences of being caught, and the student's perceptions of the teaching staff. One of the best studied areas is that of cheating in assignments, tests, quizzes and exams (Forsyth 1985; Eisenberger and Shank 1985; Radcliffe and Rudolph 2007). It is very clear from this that many students will cheat and break rules. In our experience, widespread cheating will cause even conscientious students to cheat because as we have been told "everyone else is getting the benefit so why should I miss out?"

Rules about smartphones are even more difficult for the student body to accept than rules about tests and exams. Personal use of smartphones is seen as a victimless "crime" and no one is hurt except perhaps the user and so the motivation to obey rules on smartphone use is even further diminished.

Given this background what strategies can teachers use to make any rule about smartphones? The following list of ideas comes from our experience and those of our close peers in the STEM area.

- Document any rules and discuss these rules with the students very early in the education process.
- Have a olid reasons for a rule that make sense from the student perspective. This may include not distracting fellow students, trying to optimize all students' marks, and helping students be better prepared for tests and exams.
- Explain the need to work through boring material and times in order to capture the valuable bits.
- Discuss delayed gratification: the ability to work through the boring things now to get a benefit later on.
- The consequences can be either externally applied (for example ejection from a test) or internalized to the student. For example, "I will not take your phone but I will remind you that your marks may suffer.".
- Consequences must be clearly explained and must be applied fairly and consistently. Inconsistency will gain student anger, failure to apply consequences is seen as a sign that students can "get away with it".

References

App Stores: Number of Apps in Leading App Stores 2019. (n.d.). Retrieved July 30, 2019, from Statista website: https://www.statista.com/statistics/276623/number-of-apps-available-in-leading-app-stores/.

Beland, L. P., & Murphy, R. (2016). Ill communication: Technology, distraction & student performance. *Labour Economics, 41,* 61–76.

Cell-Site Simulators/IMSI Catchers. (2017, August 28). Retrieved July 8, 2019, from Electronic Frontier Foundation website: https://www.eff.org/pages/cell-site-simulatorsimsi-catchers.

Eisenberger, R., & Shank, D. M. (1985). Personal work ethic and effort training affect cheating. *Journal of Personality and Social Psychology, 49*(2), 520.

Flam, H., Marcus, C., & Carson, M. (Eds.). (2008). *Rule systems theory: Applications and explorations*. Peter Lang.

Forsyth, D. R., Pope, W. R., & McMillan, J. H. (1985). Students' reactions after cheating: An attributional analysis. *Contemporary Educational Psychology*, *10*(1), 72–82. https://doi.org/10.1016/0361-476X(85)90007-4.

Kumar, D. K., & Radcliffe, P. J. (2019). Teaching while surrounded by smartphones. *In 2019 41st Annual International Conference of the IEEE Engineering in Medicine and Biology Society (EMBC), IEEE* (pp. 4194–4197).

Ledsom, A. (2019). The mobile phone ban In French schools, one year on. Would it work elsewhere? Retrieved November 9, 2019, from Forbes website: https://www.forbes.com/sites/alexledsom/2019/08/30/the-mobile-phone-ban-in-french-schoolsone-year-on-would-it-work-elsewhere/.

Matyszczyk, C. (n.d.). *Science teacher suspended for using jammer to shut up students' cell phones*. Retrieved July 8, 2019, from CNET website: https://www.cnet.com/news/science-teacher-suspended-for-using-jammer-to-shut-up-students-cell-phones/.

Plagiarism: Facts & Stats—Plagiarism.org. (n.d.). Retrieved July 13, 2019, from http://www.plagiarism.org/article/plagiarism-facts-and-stats.

Radcliffe, P. J., & Rudolph, H. (2007). Plagiarism: A survival strategy or recipe for disaster?

Chapter 5
Measuring the Effects

This chapter addresses the issue of how to measure the education process in relation to smartphones.

Many people have strong opinions as to what works and does not work in a classroom. There are number of reasons why this can happen such as their personal experiences, which may be biased. People may give extra weight to what they know, the current views or fads in the profession, what gets published, and discount concepts that are unfamiliar or require change in behaviour. The opposite may be true, people may be fascinated by the novelty of new and unproven technology and use it to the detriment of the education process. The bias may also be based on the author of the new concept, where a famous name may lend itself to a positive response without checking of the facts. To avoid these problems, or at least limit their effects, it is important that the new teaching methods is carefully evaluated and measured. Quantifiable measurements can remove personal opinion, can be used to monitor the progress, and can be used to convince your peers and managers as to the efficacy of the method.

One of the most interesting studies in biased educational design is the rejection of phonetic reading in the UK, America, and Australia. The experimental evidence consistently showed that learning to read by phonetics was one of the most successful learning techniques. It should not be the only technique a student uses but it was the most useful. Despite this evidence educational "experts" in the 1980s rejected phonetic reading and banned it's use in state schools. As a result, reading skills deteriorated over several decades. At the university level we found students were unable to pronounce new words they had not seen before. Eventually the weight of evidence has brought back phonetic reading, but damage had been done to generations of readers (Rowe, 2005; Rose, 2006). Imagine you have tried a new educational initiative using smartphones and announce in a staff meeting that it has been a success. There will be someone who asks the question "where is the evidence". Without some form of objective measurement, the discussion can easily descend into clashing opinions and heated discussion. If you do have objective results, then you have

D. K. Kumar and P. Radcliffe, *Teaching Surrounded by Smart Phones*,
SpringerBriefs in Applied Sciences and Technology,
https://doi.org/10.1007/978-981-15-1401-2_5

significant evidence to forestall such an outcome and you are much more likely to convince your peers that you have had success.

The difficulty in measuring the effect of a new way of teaching is—from whom should the effect be measured—the teacher, the students, the University, or some kind of weighted average. Just like the teacher may be biased, the same also applies to the students in the class who may also be influenced by laziness, or rhetoric from academics or peers, and their feedback may lack objectivity. Many Universities attempt to provide student centred learning and base a large emphasis on the student feedback. While this has some benefits, it is open to abuse and can open the doors for an extremely lenient classroom where the lecturer is unable to push the students outside their comfort zones for fear of receiving poor teaching reviews. Most academics will have seen students with low motivation give positive feedback on lax and easy education and punish classes which are harder and make them more work ready.

Another big problem is, "what to measure?" There are number of factors that can be easily quantified such as cost per class, pass rate, attendance, and average marks. Student feedback has already been mentioned. Such measurements may be what we call MIPS—Meaningless Indicators of Performance. MIPS have no direct bearing on the real goals of education which often include employment rate and skills and knowledge that industry desire. While these real goals can be hard to measure, especially in the early years of a degree, there can often be secondary indicators which can be measured. For example, in the engineering field employers nearly always rate communications skills and teamwork skills as an absolute key graduate requirement. In the university environment it is possible to both automatically and by human eye report on writing quality. Presentation style can also be measured. While these measurements may not exactly match the needs of industry, they are close (a secondary indicator) and a good indicator of the skills sought by industry. Sadly, few universities make a concerted effort to evaluate students in this area and provide feedback to guide student improvement.

Measurement in general is beset by a huge range of problems that we can only touch on in this chapter. The Hawthorn effect is a change in the behaviour of the subjects of a study, due to their awareness of being observed. In one of our early attempts to improve the teaching of electronics we brought in on-line material and discussed this with the students, the student survey results skyrocketed to very high values. The next year the same course got much lower results, the first cohort valued the newness of the initiative and the novelty, to the second cohort there was nothing new or exciting. Researcher bias can be quite accidental and so result in unreliable measurements. Much of science demands double blind trials to eliminate exactly this problem but this can be very difficult to implement in education. Confounding factors and other changes can also destroy the validity of measurements.

In order to be useful, any measurement should be the result of individual student effort. Measurements which are based on groups can become badly biased due to issues such as dominance by some group members. If copying and plagiarism is possible then the educator will see measurements which dramatically overestimate the skills of the student body. To judge the efficacy of an educational innovation before and after measurements are needed, or measurements on two similar groups who have different educational approaches used on them. This is the only way to have objective evidence on the performance of a new paradigm and not rely on gut feeling. Capturing this information needs planning and thought. In most cases, a previous class will have measurements that can be used. It may also be possible to try the new approach on part of a class and not the other part. There may be ethical issues if one group might be disadvantaged because of the methods used on them in comparison to different methods used on another group. There may also be an issue with students resenting that they are not part of the new initiative.

While measurement may be possible it may be a lag indicator, only available late in the process and too late to help detect problems early enough to fix those problems. The ideal measurement is a lead indicator which can be made early in the process and help guide improvement. Weekly feedback can be a lead indicator whereas the final exam is definitely a lag indicator as it happens after the education process is complete. Qualitative measurements may be made early and be useful in monitoring the process. A good example is the opinion of tutors in a weekly tutor meeting.

What to measure?

The foregoing discussion examined the many difficulties in measuring the education process. If the problems can be largely overcome what can be measured in order to compare any new educational activity to the old way of doing things?

Traditional measurements which are still of great use include:

- Tests and quizzes that are identical or nearly identical.
- Attendance rates as time progresses.
- Student surveys, with the same questions.
- Student complaints or praise, from a substantial number of students.

Qualitative feedback is not as useful as quantitative feedback as it relies on human judgement, which may be biased. Qualitative feedback is usually better than no feedback and can still guide improvements. Typical qualitative feedback includes:

- Student complaints or praise, from a small number of students.
- Tutor opinions.
- Group discussions in classes, labs, or tutorials.
- The judgement of an experienced, practicing teacher.

Technology can provide more measurements:

- Usage rate of teacher provided resources. Many Learning Management Systems (LMS) make it very easy to generate quite complex statistics about student use.
- Usage rate of apps and online tools provided by the institution. Remote desktops apps provided by many universities can provide quite detail usage statistics.
- The growth of documents or software code over time as evidence of steady work effort. This can be easily managed using tools such as GitHub.
- The individual student contribution to collaborative tools and outputs.

Usage rate measurement needs to be carefully thought through. In one of our colleagues courses the statistics on resource usage look quite good. The hit rate was about 80% of the class. Our colleague dug a little deeper and extracted the number of unique students accessing the resource. Only 15% of the student body had accessed the resource! Many had repeatedly accessed the resource, most likely due to their poor English skills.

If publishing your results is a possible aim, then take time to read the literature and see what others have done. Higher ranking journals usually have guidelines about what types of measurements must be made before the article will be considered.

And if you cannot measure?

Ideally you will be able to measure in the ways suggested, but in many cases, this is simply not possible. There may be no relevant historical data available as courses have been changed or are new, or there are confounding factors. In nearly all cases there will be some qualitative feedback that can provide some idea of the progress of an activity. This information may be available from discussions with students and tutors, comments in student surveys, and observations of student behaviour. Such qualitative information is less reliable than quantitative measures but better than no measurement at all.

Conclusion

Measurement of an educational activity is difficult, requires effort, and requires considerable thought and planning. There are numerous problems and confounding factors and in some cases measurement may simply not be possible. If measurements have been made, then you have some major ammunition to convince others of your successes and ensure the "idea hitmen" cannot belittle or sideline your achievements. The right type of measurements may enable you to publish your experiences and so contribute to the body of knowledge.

References

Rose, J. (2006). *Independent review of the teaching of early reading: Final report*. Retrieved June 24, 2019, from http://dera.ioe.ac.uk/5551/3/5d970d28fc535dc54eb4dee9995bef36.pdf.

Rowe, K. (2005). National inquiry into the teaching of literacy (Australia), & teaching reading: Report and recommendations. *Teaching and Learning and Leadership*. Retrieved from https://research.acer.edu.au/tll_misc/5.

Chapter 6
Case Studies

This chapter provides a number of case studies that highlight the strengths and weakness of different approaches introduced in earlier chapters. All these case studies were intended to improve the learning outcomes for students when the students have access to their smartphones or other similar devices. Not all case studies were successful but, in all cases, important lessons were learned. We know that there will be many more approaches people have used to cope with smartphones, many of which have been very successful. The case studies described here are representative examples drawn from our own experiences and those of close peers in the STEM area (Kumar and Radcliffe 2019). We have found that discussion of case studies engages academics and starts a conversation on how best to manage smartphones in education. We hope you can start such a conversation with your peers by sharing the case studies presented here and your own experiences.

It is very important to report the context of each case study because this context can dramatically affect the outcomes. What will work in one cultural environment may fail dismally into another. We have found a tremendous difference between highly motivated students and those with low motivation which is in line with the literature (Beland and Murphy 2016). What worked well with one group did not work in another group. The problem gets even more difficult in classes with a large range of important attributes. For example, in one class, students from some degree programs can be very interested in a topic, but those from other degree programs are deeply bored.

Do not attempt to blindly reuse any successful methods presented here and expect success. You will need to compare the context of the case study to your own and try to determine if this context is close enough to your own. All case studies have lessons that might be useful and might be extrapolated to your environment.

D. K. Kumar and P. Radcliffe, *Teaching Surrounded by Smart Phones*,
SpringerBriefs in Applied Sciences and Technology,
https://doi.org/10.1007/978-981-15-1401-2_6

6.1 Case Study 1: Policing the Use of Smartphones—Large Classrooms

Aim

The academic had found that the students were getting very distracted and not paying attention to the lesson. Many of them were very openly using their smartphones, and the biggest problem was when some of the phones rang loudly, which disturbed the rest of the class. The academic attempted to clamp down on this by using authority and banned the students from keeping their phones in on-state while in the class.

Background of the student group relevant to this case study

1. Second year university students in a mid-tier university with an average age of 19 years.
2. Students from Science and Engineering and are expected to be well-versed in technology.
3. Most of the students come from across the state of Victoria, Australia. Most students would know only a few colleagues in the class.
4. Around 20% of the students are from overseas.
5. Around 20% of students were females and 80% males.

Background of the class management

- It was not compulsory for the students to attend the class and no attendance was taken. While the enrolment numbers were in the range of 200, typical attendance was between 50 and 100 students in the class.
- The room was well equipped and modern, and designed for good accessibility between rows. The podium was located in the well of the theatre. The seating was for 200 students, with a horse-shoe arrangement and split in three columns.
- The lecture theatre was designed for ease of movement from the front row to the back of the room, and there was one entrance from the back and one from the front of the room.
- In this class, the students were not expected to use their laptops, though they could use these to take notes.
- The room had the provision of wireless Internet and power-point for each student.
- The podium provided the lecturer with control of the audio-visual and lighting. Six wireless microphones were provided such that the lecturer may distribute these to the students for facilitating discussions or for students to ask any question.

Relevant Action

In the previous classes, the lecturer had been very disappointed because the phones of the students rang many times which was disturbing to the other students. He was keen to prevent the repeat and announced at the start of the lecture that all students should switch-off the phones and keep these in their bags. The lecturer walked around the class, and repeatedly reminded the students to ensure that the phones were not on their person. The students were threatened with undefined punitive actions if they did not comply.

A majority of the students kept their laptops on their study desks in front of them. The lecturer reminded the students not to use their laptops during the lecture, the request which was ignored by some of the students.

The class was scheduled for two consecutive lectures, each being 50 min. The lecturer took the two slots without a break. The lecture was based on well-developed power-point slides with an embedded short video. The lecturer was an experienced teacher and made a significant effort to make the class interactive by raising a number of questions and asking the students to work in pairs to find a solution.

Unknown to the students, at the back of the class sat a teaching assistant who monitored the class on behalf of the lecturer. Her seat was located in the highest row of the room, and she could observe all students' activities, even if happening in their laps. She made notes when she could notice the students using their phones.

Discussion and Results

There was only one event when the phone ringing sound was heard in the room, and when this happened, the student quickly stopped it from ringing.

The observer sitting at the back of the room noticed that many of the students did attend to their phones quite regularly- perhaps to check their emails or messages. The phones were typically placed on their laps or hidden behind papers placed on their desk.

From a distance, it was observed that many of the students did have their laptops open despite the warning from the lecturer. In most cases, it was difficult to judge whether the students were using these to take notes or do non-classroom activities, though in a small handful it was evident that the laptops were being used for entertainment purposes.

This study shows that most students in the class are respectful to the extent that they did not allow their phones to ring loudly and be heard. This is important because the ringing of the phone disturbs other students and the teacher, and it is essential that this is avoided. From this viewpoint, the ban of the phones may be considered to have been effective.

The study also shows that students did not consider switching-off the laptop important even though this was asked for by the academic. Perhaps this was because the general expectation in the class was that laptops were permitted and did not disturb other students, so the students did not take the request seriously.

It is also seen it was not possible for the academics and teachers to stop the students from keeping their phones on them and from using these discretely. While this may have achieved a purpose, it does however cause the students to be distracted and not pay attention to the lecture. Another difficulty is that the students sitting close to the one who is attending to the phone are also distracted.

This shows that the academic was unable to prevent the university students in year two from paying attention to their phones while sitting in large classrooms. While there can be a number of reasons for this, one simple observation is the room itself; the size and shape of the room does not allow the teacher to view the desk of the students. Thus, the students appear to connect with their phones while not expecting the teacher to notice that.

There are other possible reasons for this distraction such as the size of the class and type of students. When there are large number of students, it is difficult for the teacher to observe individual students and they may get an impression that the teacher will not notice. The other factor may also be that the students were not the high-achievers and lacked self-motivation, making them more likely to be distracted. While it is difficult to identify the specific reason for this failure, this study clearly shows that attempting to police the students to keep their phones with them in a large classroom does not work effectively.

6.2 Case Study 2: Policing the Use of Smartphones—Small Classrooms

Aim

This class ban on mobile phones was a reaction to an event where a smartphone rang loudly during the classroom and disturbed the entire class. The obvious aim of this ban was to prevent the class from getting disturbed and the method was to use the teacher's authority to ban the smartphones from being used in the classrooms.

Background of the student group relevant to this case study

1. Third year university students in a very high-end university. Average age 19 years.
2. The students come from across India and reside in the dormitories provided by the University and all students know each other very well.
3. Around 30% students were females and 70% were males.

Background of the class management

- It is compulsory for the students to attend the class and attendance is recorded. The enrolment numbers were 40, and typical attendance was around 38 students.
- The classroom was a flat room with 10 rows and 2 columns, each column having two students.
- The room had wireless Internet and power-point for each student.
- There was no formal podium. A computer and audio-visual control were at the front of the room.

Relevant Action

The lecturer had been very disappointed because the students had been using their smartphones during the previous lectures. While no student phone rang loudly, however the lecturer was unhappy because some of the students appeared to be distracted because of their phones and were not paying full attention to the lecture.

The lecture announced at the start of the lecture that all mobile phones had to be switched off and placed on the study desks in front of the students. The lecturer then walked and randomly selected a few of these and checked if these were switched off.

The class was scheduled for 55 min and the lecturer was using power-point slides. There were a number of questions asked by the students and the class was very interactive.

This style of the lecture was repeated for the entire semester. At the end of the semester, a simple feedback was taken from the students to evaluate the following questions:

1. Did they believe that all (other) students did switch-off their phones?
2. Did they believe that this was good for the classroom?
3. Did they believe that they will be failed if they were caught with their phones turned on?

Discussion and Results

One observation by all the lecturers was that the ringing of the phone in the classroom stopped in entirety for the semester.

The feedback form suggested that most students (>90%) believed that all phones were switched off. The feedback also showed that they did feel that having the phones switched off was a good thing for the class and for their learning. It was also evident that the students were convinced that the academic had the authority to fail them if their phone was switched on, that is, they believed that the academic had full authority. However, a number of the students did complain that this was draconian and not an appropriate method.

Banning smartphones is a method to attempt to return to old-fashioned classrooms. This approach can result in reduced distraction for the students but will only work when the students are high-achievers or highly motivated, and the University is a highly sought-after destination which does fail nonperforming students. It is important that the rooms are small such that the academic can have all the students in the visual field. But perhaps the most important aspect was the sense of authority, that the academic had unquestionable authority and the students would not dare to disobey. While this method can be successful in preventing the distraction due to the smartphones, it does not use the devices to the advantage of the learning experience.

6.3 Case Study 3: Sharing the Teaching Material On-Line in the Class

Aim

The aim of this teaching innovation was to improve the learning outcomes of the students and reduce distractions in the class by harnessing smartphones to benefit the educational activity and thus stopping distracting use of those phones.

Background of the student group relevant to this case study

University students in year three from Engineering discipline. Class size of approximately 100.

1. The students could be referred to as mid-range, having entered the University from high school with score at or above the 80th percentile, in Melbourne, Australia.
2. Most of the students come from across the state of Victoria, Australia. Most students would know only a few colleagues in the class.
3. Around 20% of the students are from overseas.
4. The demographics of these students were mixed, and both genders were represented, though there were considerably more male students. The typical class had around 20% female population.

Background of the class management

1. It is not compulsory for the students to attend the class and no attendance was taken. While the enrolment numbers are in the range of 200, typical attendance is around 50 students in the class.
2. The room is designed for good accessibility between rows. The podium is located in the well of the theatre. The seating is for 200 students, with a horse-shoe arrangement and split into three columns.
3. The lecture theatre is designed for ease of movement from the front row to the back of the room and with entrance from the back and the front of the room.
4. The room has the provision of wireless Internet and power-point for each student.
5. The podium provides the lecturer with control of the audio-visual and lighting. Six wireless microphones are provided such that the lecturer may distribute this to the students if there is a discussion or question.
6. An observer was seated discreetly near the back of the class to see to the extent students used their smartphones.

Relevant Action

The lecturer uploaded a copy of the lecture material on the web and shared the URI with the students. The lecturer did not project the slides on the screen in the class but asked the students to see the screen on their personal devices.

Discussion and Results

The activity was quickly considered to be a disaster because of the large number of complaints and so the usage of smartphones was discontinued. As a result, it was not possible to perform any measurements. The only positive was that all the students had their personal devices and the wireless Internet in the lecture theatre had sufficient capacity for all the students to connect to the Internet effectively.

The most rampant complaint was from fellow students complaining of the disturbance being caused by the devices of other students, particularly those using their phones for non-course related activities. Another significant issue was the number of students who could not find the website causing them confusion and frustration. The biggest cause of complaint came from the lecturer himself when he was confronted by a single case of bad behaviour; one student began to celebrate very loudly while watching a soccer game during the class.

The use of smartphones and other portable digital devices in this class did not deliver the desired outcomes and the students were distracted because of the devices.

Below are some of the points that attempt to explain why this teaching experiment was not very successful:

1. The students were largely left to themselves to determine the role of the devices in the learning. The students had been informed by the lecturer to use their personal devices to link to a website which contained additional material such as demonstrations and examples that would help the students better understand the topic. This was in addition to the lecture material that was being provided by the lecturer. However, the lecturer did not use any method to direct or monitor the students, or to engage with them, and this resulted in students getting distracted away from the lecture material.
2. The activity required a degree of multi-tasking that was beyond many students. Should they listen now, or go to the web site, or give up and watch the soccer.
3. In a traditional class-setting, the lecturer is able to observe the students to understand their progress and get their feedback, but in this situation the students were busy looking at their screens. The lecturer was giving a lecture, the students were expected to have the devices on their desks, and this made it very difficult for the lecturer to observe and understand the level of engagement of the students with the material.
4. In this class, attendance is not compulsory, and no attendance records are maintained. Thus, it might be assumed that the students are attending the class because they are keen and engaged. Coming recently from high school they may just be in the habit of attending classes in body, if not mind. However, the observations suggest that while a significant number of students were attending the class because they wanted to listen to the lecture, they seem to have got mixed messages when they were encouraged to use their phones or tablets to get the lecture material. It is possible that if this style of lecture was repeated a number of times, the students will get used to the idea of seeing the material on their device rather than just listening to the lecturer.
5. Another shortcoming in this approach was the students, who can be considered to be in the middle-range of academic excellence and in the age bracket around 19 or 20 years. Such students are not expected to be the high achievers as is evident from the university entrance scores. While such students have been found to perform when directed, they are also prone to distractions. The typical student in such a cohort would have lacked high-level self-motivation which would be essential for the students to remain focused on both the lecture and the material and avoid the distractions provided by the digital devices.

Other comments

The was significant effort required on the part of the lecturer to prepare material for the smartphone, in addition to the lecture material which was prepared as a series of power-point slides. Just these activities alone required significant preparation effort and required a large time commitment from the lecturer. In this case, the lecturer also had prepared examples to help the students better understand the slides and this had taken him many hours of planning and preparation. The result was worse than a standard lecture, far more effort, and the whole episode felt very disappointing.

6.4 Case Study 4: Using Purpose Developed on-Line Material

Aim

The aim of this study was to improve the learning outcomes and reduce the distraction of the students due to their phones and social media during the lecture. Another aim was to make the classrooms more engaging for the students to ensure a higher retention rate of the educational material. It was found that the students attend to their personal digital devices when in the classroom listening to the lecture. This results in these students being unable to give quality attention to that material. While they generally have the phones on silent mode, this activity was disturbing to the other students and the lecturer.

Background of the student group relevant to this case study

University students in year three from an Engineering discipline. Class size of approximately 100.

1. Most of the students come from across the state of Victoria, Australia. Most students would know only a few colleagues in the class.
2. Around 20% of the students are from overseas.
3. The demographics of the group were quite mixed and around 20% of the class were female.
4. The students could be referred to as mid-range, having entered the University from high school with score at or above the 80th percentile, in Melbourne, Australia.

Background of the class management

1. The class is a tutorial with a flat floor room, white board, and audio-visual equipment at the front of the room.
2. It is not compulsory for the students to attend the class and no attendance was taken.
3. There is one tutor and a class of 20, but the typical attendance is only 14 students.

Relevant Action

In this example, the students were expected to use their personal digital devices to solve the given engineering problems. The tutorial was in a tutorial room which was flat style. The tutor presented a series of problems which the students worked through using their smartphones. At the end of the session there was an online test that was automatically marked, and students received the results after answering each question.

Results and Discussion

Some of the major outcomes of this study were:

1. One of the significant improvements was the feedback from the students who believed that they were able to understand how to solve the problems in the classroom itself. The student feedback score was found to improve significantly.
2. The students were very positive about receiving immediate feedback via the online quiz at the end of the tutorial and this was reflected in the student survey at the end of the course. Students felt it gave them a good indication if they were performing well or needed to work harder on a particular area of weakness.

This is an example where the outcomes of managing the classroom when all the students had personal digital devices were positive. The feedback from the students was positive and the tutor was satisfied with the outcomes. This teaching method shows that students are able to use their digital devices effectively when they are given directions for specific tasks that have to be done. When the students had to use the smartphone for performing the specific analysis and solve the specific problems, they were able to do this very well, did not get distracted by other things on their smartphone, and enjoyed a positive learning experience.

The other major benefit the students experienced was that this approach had given them an opportunity to evaluate their knowledge while in the presence of the tutor. The availability of their smartphones allowed them to perform self-evaluation immediately at the end of the class. This allowed them to clarify any doubts and improve their understanding of the topic, and this enhanced their learning experience. They appeared to have treated this evaluation as a positive approach, especially because the results were not being recorded but were only for their personal satisfaction.

A significant issue with this approach was that the tutor had to make a significant effort in developing the resources for the students. This extra labour would need to be paid for if the technique was more widely adopted. Despite all this effort, the number of students attending the tutorial did not increase.

6.5 Case Study 5: Using Clickers in Class

Aim

Clickers are devices that allow students to give immediate feedback in class. The idea behind their use in the classroom is that such devices encourage student participation and make the lecture more attractive than their smartphones. There are several clicker styles available for smartphones. This case study will show that clicker usage is fraught with issues and that careful forward planning is required if these are to be effectively used to improve the learning outcomes for the students.

Background of the student group relevant to this case-study

University students in year three from an Engineering discipline. Class size was approximately 300.

1. Most of the students come from across the state of Victoria, Australia. Most students would know only a few colleagues in the class.
2. Around 20% of the students are from overseas.
3. The demographics of the group were quite mixed, around 20% of the class were female.
4. The students could be referred to as mid-range, having entered the University from high school with score at or above the 80th percentile, in Melbourne, Australia.

Background of the class management

1. Attendance is not compulsory, and no attendance is recorded.
2. While the enrolment numbers are in the range of 300, typical attendance is around 150 students.
3. The room is a refurbished cinema with good accessibility between rows. The light levels are low making reading a little difficult. The seating is for 300 students with straight rows facing the podium.
4. The room has no wireless Internet and no power-points for students. However, phones work inside the theatre and the Internet is available through the phone system.
5. The podium provides the lecturer with control of the audio-visual and lighting. It is located on a stage such that a few rows are below the podium with the rest above.
6. The lecturer had 8 clickers which were distributed to the student audience. These clickers could register several options and the total results could be seen on the lecture screen.

Relevant Actions

The aim of the activity was to engage students in the lecture material by posing questions throughout the lecture. The general concept behind this design was that the clickers will help improve the engagement of the students with the class educational activities and thus improve the learning outcomes.

According to the instructions, students with the clicker would respond to the question and the results will be displayed on the lecture theatre screen. Given the limited number of clickers that were available, the lecturer planned to ask students to work as a group of 6-8 students in order to get the score and decide what answer to give. The lecturer planned to walk around the lecture theatre and handed out the clickers to each area of the theatre.

The proposed method of asking questions was varied. One approach planned was to ask a question before a body of theory was discussed and then gain student opinion via the clickers. The theory was then delivered, and the same question asked again to determine the successful training received by the students. A more basic approach could have been to simply ask some fundamental question to test if theory that had been presented had been understood by these students. In this course, there were no calculations that were appropriate but in other courses it may be appropriate to pose numerical questions that could evaluate the learning outcomes. Finally, all the theory was packaged into 10-min segments and questions were written for each short segment. A five minutes period was allowed to pose a question and get the student responses.

Discussion and Results

The clicker driven lectures were run three times, spread over three weeks with the same group of students. The start of the first lecture successfully explain the use of the clickers, handed them out, and ran a trial question and answer. There was initial enthusiasm for this new technology for the first three or four questions, but after that it was noticeable that in some groups one person was dominating the decision making and the other students were being ignored, or possibly had lost interest. After the first hour less than half the groups were making group decisions as to the answers. The behaviour of students without a clicker was also interesting. Many students went back to their smartphones and were not engaged with the lecture. At the end of the lecture, the lecturer asked for the clickers to be returned, two were not returned and could not be traced.

The second and third lecture used clickers but the acceptance from students was poor with many not bothering to register an answer. The five minutes allowed for student to formulate and answer appeared to break student's concentration and caused them to be diverted onto other activities such as using their smartphone. Clearly the student acceptance was low and after the third lecture clickers were abandoned. Lectures after this did not use clickers.

The attempt to use clickers had issues primarily with the planning of the class. The planning phase of any educational activity must seek to predict the behaviour of students and to optimize the engagement in the education activity. The use of only eight clickers for an attendance of approximately 150 students is clearly problematic. Even when the clicker is used by a group this leaves many students out of any action and so they become observers and less engaged. The random allocation to groups of students who may not know each other is fraught with problems. Most likely there will not be group consultation when students do not know each other. There is also a strong chance that a dominant or skilled student will simply take over the operation of the clicker and leave the other students as observers, who may resent being relegated to observers. In order for a clicker system to engage the student body rather more than 8 clickers for 150 students is required.

Another important planning issue is how to ensure that clickers are not lost or pilfered. This was not thought through and two of eight clickers were lost in the first lecture, and only 3 out of 8 were recovered by the end of the third lecture. Solutions here might include clear labelling with the universities name and recording who was given each clicker.

The observation of the execution raised some interesting issues. The initial use was promising with good student engagement, but this engagement wore off after a short while. There was clearly interest in this new approach but after a while the novelty wore off and the clicker system did not engage many students and after a while many students returned to their mobile phones. Once the novelty has worn off is a good time to judge the utility of a methodology and in this case the methodology did not help the education process.

The formulation of engaging questions for the clicker is much more difficult than one might suppose. The easiest questions to pose are memory questions but these do little to engage students in real learning.

The use of clickers as stand-alone devices or even on smartphones must be questioned. A standard part of any planning must be to consider alternative methods of achieving a similar result. Software such as Kahoot can be run on a student's mobile phone and display on the lecturer's web browser window. This is more powerful and engaging than a clicker and is available to everyone with a mobile device.

Key conclusions

- Any class interaction system must engage all students, not just a subset.
- There are many alternative ways to interact with students in a lecture scenario and a wide selection should be considered before choosing one.
- Planning before an activity should identify how each student type or subgroup might react. Additionally, a careful "what could go wrong" analysis should take place.

6.6 Case Study 6: The Flipped Classroom and Smartphones

Aim

The flipped classroom is a well-known approach to improve student engagement. Students should have read some material and completed some exercises before a lecture. The classroom is then changed from being a lecture to a question and answer session, possibly with more exercises and with small components of problem-based learning introduced.

In the case study being reported here, a flipped classroom approach was tried. However, the results were not good; in fact it was a failure. The approach failed dismally in part because mobile devices distracted the students with more interesting material they found on the Internet, or they interacted between themselves on topics irrelevant to the topic being taught. The aim of sharing this case-study aims to help identify the flaws in the approach and issues that need to be resolved before a flipped classroom approach can be successful.

Background of the student group relevant to this case study

University students in year 3 from an Engineering discipline.

1. Most of the students come from across the state of Victoria, Australia. Most students would know only a few colleagues in the class.
2. Around 20% of the students are from overseas.
3. The demographics of the group were quite mixed, around 20% of the class were female.
4. The students could be referred to as mid-range, having entered the University from high school with score at or above the 80th percentile, in Melbourne, Australia.

Background of the class management

1. This class was a tutorial intended to clarify lecture material, but attendance was not compulsory.
2. The tutorial group size was approximately 24 students.
3. The room was arranged with seven round tables that could take 6 students on each table. The tables facilitate student group discussion.
4. The room has wireless Internet and power-points for students.
5. There are multiple TV screens around the room so every student can see a screen, though they may have to turn their seat around to see a nearby screen.
6. The podium provides the lecturer with control of the audio-visual and lighting. Many students have their back to the podium due to the round tables.

Relevant Action

The university is strongly in favour of the flipped classroom approach and has built the tutorial rooms specifically with this in mind and has equipped the rooms with multiple audio-visual facilities. The flipped classroom approach was new to the target student group and so the approach was carefully explained in the first class. The expectations on the student body were clearly spelt out; that they would familiarise themselves with the material before the class. They were specifically told to read the notes before the lecture and to attempt some of the practice problems which had been provided. Students were provided with notes and a reading list and some practice problems before the session. During the tutorial the lecturer planned to ask questions to test the student's understanding of the material and work through more problems to help the learning of the students.

Discussion and Results

The tutorial session started with the lecturer asking questions based on the material the students should have read. A small group of students responded, and the lecturer then attempted to involve the other students and asked students who did not volunteer an answer to make a contribute. It was evident that the majority (around 75%) of the student body had done none of the preparation work and many of this group started using their smartphones rather than participate in the session. The response of students who had not read the preparation material varied, some were embarrassed, some were resentful at being picked out, while some were downright annoyed that they should be expected to work before a tutorial.

The tenor of the session was quite negative, and the expected learning outcomes were not achieved. The second session resulted in a significant drop in attendance and for those who did attend, the use of mobile devices to pursue non-academic and other activities increased and was no longer discrete. The flipped approach was dropped after two more sessions and the tutorial returned to the previous approach where students worked through problem sheets. In these classes the use of mobile devices dropped dramatically and moved from weaker students to strong students who had finished the work sheet early.

Probably the biggest problem in this case study was a failure to model student behaviour in the planning phase. Students were used to a tutorial with problem sheets and a step by step guidance from the tutor. In this existing culture, no preparation was required and there were no consequences for not preparing. Despite the flipped classroom approach being explained there was no incentive for students to change their habits and actual do the preparation work. Changing culture often requires an extra incentive to encourage people to make that change. For example, what would have happened if the tutorial had started with a brief test that carried some marks, a consequence that could have mattered? Would the students then have been incentivized to complete the preparatory work?

Smartphones contributed to the demise of the flipped classroom experiment because they provided a way out for students who were bored or perhaps embarrassed by their lack of preparation. There was no incentive to pay attention or contribute when email or social media was easily available. It would be interesting to see what students would have done if their mobile devices were not available, but realistically such an experiment would not be possible.

Key conclusions

- New educational approaches must first be clearly explained to students. But this is not sufficient, and they must also be given an incentive to change behaviour.
- Majority of these students were clearly not highly motivated and so did not do the extra work to fit into the new paradigm.
- Dealing with mid-ranking students may require a carrot and stick approach, with stricter monitoring and more direction than could have been provided.
- A large part of the student body from a middle-ranked cohort may not participate if there are no consequences.

- Smartphones provide an easy way for students to mentally opt out of an educational activity, and thus the potential positive outcomes become negative.

6.7 Case Study 7: Using Kahoot in Class

Aim

A number of academics have reported the use of smartphone apps and tools for improving the learning outcome of the students. One of the popular such tool reported in recent literature is Kahoot which aims to make the classrooms more interactive and more engaging for the students. This activity aimed to test this observation.

Kahoot is a tool that captures student opinion via a web browser on any Internet connected device. The class results can be viewed on the lecturer's web browser which can be shown on the lecture hall audio-visual system. Kahoot has many other features and can be used for free in schools or for personal use though the full power of Kahoot requires paid access. See https://kahoot.com/welcomeback/.

Background of the student group relevant to this case study

The class was approximately 300 university students in year 1 from an Engineering discipline.

1. Most of the students come from across the state of Victoria, Australia. Most students would know only a few colleagues in the class.
2. Around 20% of the students are from overseas.
3. The demographics of the group were quite mixed, around 20% of the class were female.
4. The students could be referred to as mid-range, having entered the University from high school with score at or above the 80th percentile, in Melbourne, Australia.

Background of the class management

1. Attendance is not compulsory, and no attendance was recorded.
2. While the enrolment numbers were approximately 280, typical attendance is around 180 students.
3. The room was a purpose-built lecture theatre where the student seats for an arc focused on the podium. The seats are raked at an angle which ensured good visual and audio contact with the podium.
4. The room has wireless Internet and power-points for students.

Relevant Actions

In the first year of a university degree, it can be very hard to get students to act on issues related to their own employability. Some students are very mature and latch on to the material immediately and change the way they behave now to ensure greater employability in the future. Many students see employment as a long time away and that other things are far more important. There is also a problem in that students still adhere to the high school mentality that marks are all that matter. There are a large group who naively believe that “Ps make degrees”, that a pass degree will be enough, and a pass degree will be a meal ticket. The employability lecture aims to inform students of the many advantages of driving their own employability agenda, and the dangers of doing nothing and just thinking about marks. The employability message must appeal both at an intellectual level and an emotional level if it is to succeed in changing student behaviour, in particular taking on more effort and work now for later gain. This is a real test of the student’s ability to cope with delayed gratification.

Kahoot was chosen as a way to get high levels of student involvement and to obtain results in a visual, textural, and audio manner. The goal was also to help many students see they were wrong in their assumptions about employability and that they would fall behind students who understood the problem more clearly and acted on that understanding.

The lecture first introduced the topic and quickly moved into asking questions using the Kahoot system. Most questions were asking students to predict real attitudes from employers. The collated student opinions were then displayed, and finally the results from employer surveys. For example, students were asked to predict what the employment rate of engineers was for current graduates. Each student would log onto the Kahoot web URI indicated on the lecture hall screen and add their opinion. They were also asked to tell their neighbour what they had chosen and why. There was a very high level of student engagement and discussion. The real employment results were then shown on the screen, the lecturer talked about the results, and asked for questions from the student body. This cycle was repeated for topics related to what employers hope to see in graduates and the most effective ways to build the skills and attributes employers expect.

The final question asked students for a 120-character list of key activities they would pursue to improve their employability. With some 180 students the message flew past quite quickly on the lecturer’s screen which is visible to all. This speed was fortunate as a number of the messages were very inappropriate but mercifully these scrolled away quite quickly.

Discussion and Results

The lecturers were quite skilled and worked very effectively on the carrot-and-stick approach. For example, students were asked to predict the salary of a graduate engineer and then a retail shop assistant. This had the effect of making an engineering career look promising and the alternatives far less attractive. The cycle of asking students to predict and then showing the real situation worked quite well and did turn the lecture into a game. Asking students to discuss their prediction with their neighbour also increased the appeal of the session and most likely helped students remember the information. The topic of what employers expect in a graduate was clearly a surprise to many students who had previously assumed that high marks were the key to employment. In the engineering field the data is clear that employers see written and oral communication skills as most important followed by teamwork and then technical skills. Additionally, the employer wants high skills in all of these attributes not just some accumulate 50% pass. The impact on students was much higher using the question-predict-answer format than simply presenting the results.

The lecture finally turned to activities which could help a student improve their employability. The final question required a text response and as discussed there were several inappropriate comments. This highlights a general issue with responses that cannot be tracked to an individual, it is easy to make inappropriate comments and so some students will do so. Even tracked responses are a problem as some students will still be inappropriate requiring staff to identify and counsel the individual which is an extra workload. It is generally better to avoid problems rather than clearing up afterwards. With the benefit of hindsight, it would have been better to design the final question such that students selected from a list rather than have free text entry.

Right through the session, students had their mobile devices out but only paid attention to them at the appropriate time in order to interact using Kahoot. However, the number of students using their mobile phone as a distraction was much higher in the traditional lectures run in the other weeks.

Talking to the lecturers after the lecture, it was clearly a lot of work to setup the session. The appropriate questions have to be determined, the real data obtained, the lecture written, and the Kahoot system all set up and run. This is a common theme in non-traditional lectures, such approaches require a lot more staff effort, and extra time must be allowed for staff to prepare the material. Failure to allow staff this preparation time will result in a poor outcome and the abandonment of the new methodologies.

Key conclusions

- Interactive lectures facilitated by smartphones and tools such as Kahoot can increase student engagement.
- The distraction issue was actually worse than in a traditional lecture as students already had their phones in hand. Their use of the phone was far less noticeable as everyone had a phone in their hands.
- It is essential to analyse what-could-go-wrong to avoid problems and embarrassment.
- Skilled lecturers are required to ensure these sessions are run effectively.
- Interactive lectures using smartphones requires extra time and effort to prepare.

6.8 Case Study 8: Google Forms

Aim

One lecturer who had used Kahoot was unhappy because it required post-processing of student input during the lecture. The academic trialed using the venerable Google forms to get information from students and into a spreadsheet. Spreadsheet functions including complex graphing was used to make results immediately available to the lecture group.

Background of the student group relevant to this case study

University students were in year two from an Engineering discipline. Class size of approximately 150.

1. Most of the students come from across the state of Victoria, Australia. Most students would know only a few colleagues in the class.
2. Around 20% of the students are from overseas.
3. The demographics of the group were quite mixed, around 20% of the class were female.
4. The students could be referred to as mid-range, having entered the University from high school with score at or above the 80th percentile, in Melbourne, Australia.

Background of the class management

1. Attendance is not compulsory, and no attendance is recorded.
2. While the enrolment numbers are in the range of 150, typical attendance is around 100 students.
3. The room is a purpose-built lecture theatre with steeply raked seating.
4. The room has Internet and power-points for students.
5. The podium provides the lecturer with control of the audio-visual and lighting.

Relevant Actions

The lecture was split into multiple 10-min, individual student activities, each of which started with a brief (~1 slide) commentary and instructions from the lecturer. Students were then given time to access the indicated web resources and often encouraged to discuss the topic with their neighbours in the classroom. Each student was then expected to fill in a Google form from a link emailed to them. The raw student input was not visible to other students, but immediately processed by the Google spreadsheet into collations or graphs. These were shown and discussed with the student body. Student participation could easily be measured in terms of submitted forms or correct answers.

Results and Discussions

The pattern of introduction, web activity, and results discussion took longer than anticipated and most non-trivial activities took 15 min or longer to complete. Student involvement was good though approximately 1/3 of the class did not engage with their neighbours as requested. Some of the initial questions were open ended and aimed at stimulating students learning.

The result was that many students got lost and participation became lower than without this intervention. Well directed questions with a clear aim and limited time frame seemed to better engage the lecture group, resulted in more discussion and student engagement. Basing any marks on correct answers was problematic as the answer was often the results of a group discussion and not a reliable indication of individual achievement.

The lecturer felt that this approach worked well for some topics that were opinion based, but the approach was weak at introducing a complex body of knowledge. He felt that a traditional lecture backed up by tutorial problems managed to get across greater content and could cope with more complex knowledge bases.

6.9 Case Study 9

Style: Short Question Based Learning (QBL) with Inverted Classroom:

Aim

The aim of this activity was to improve the student engagement in a course which in the past had received unfavourable feedback from the students. While there was no specific problem that had been identified, the biggest challenge was that research students had to undertake this compulsory course, and they were unable to appreciate any purpose in the existing course. Unlike all other courses in the University, this course required compulsory attendance, and this made the students even more unhappy. Anecdotally, it was observed that during the lecture, all students were busy doing other things, and were constantly on their laptops or phones, and left soon after the attendance was marked.

Background of the student group relevant to this case-study

1. The class was approximately 100 university postgraduate students doing their research degrees. The majority of students were from Engineering disciplines.
2. A significant number of students were from overseas. Most of the students did not know many of their colleagues in the class.
3. The demographics of the group were quite mixed, around 40% of the class were female.
4. The students would have joined the University after having completed their undergraduate or even postgraduate degree from universities around the globe.

Background of the class management

- Attendance was compulsory and attendance was recorded.
- The enrolment numbers were approximately 100.
- The room was a purpose-built inverted classroom, with one central podium, and 4 screens around the room. The seating was around tables and thus the students at in small groups. They could look in any direction to view the screen.
- The room did have steps resulting in good visual and audio contact with the podium and the display screens.
- The room has wireless Internet and power-points for students.

Relevant Actions

The research students are in general unhappy with attending a class where they believe they are given information that will be not useful for their research and thus their degrees. This is further compounded because their supervisors seem to agree with this view and reinforce it with their students. Thus, the audience considered the compulsory attendance to be a punishment. One of the single biggest challenges was that the lectures seemed to contain the information that was also available online, which was mainly concerned with the rules of the candidature. Most students did not believe that they would even need to know the rules and thus the lectures were a waste of time.

To overcome the problem, and to get the students interested in the course, the academic decided to run this course in an alternative style. Based on the overall aim of the course, the students were to improve their communication skills, improve their publication capability, to network with colleagues and to ensure their research was conducted ethically and under the umbrella of appropriate research governance. For this purpose, the course was run such that there were no lectures, and a classroom suitable for inverted classroom was chosen. The students were encouraged to sit in groups of 5, and with emphasis on changing the group members every session.

The academic first explained to all the students the style of the course. They were made aware that there will not be any lecture, or notes, or slides. The strategy was discussed, and a number of students got quite engaged in this discussion. The lecturer also explained the reason for the strategy, which was to ensure that the students networked, and did not have to feel guilty when using their smartphones. The aim and objectives of the course was also discussed which concluded the introduction to the course. Students were also informed that they could leave the room for a coffee break at their will. The only rules were that they should not speak on the phone and the phones should be on silent mode.

The academic then began by providing the students with small questions, and the students were asked to search on the net to find answers, after which they had to discuss their findings among their group. Subsequently, the group had to identify their script writer, who had to enter the group response into an online document which was shared among all the students in the class. This document was managed by a teaching assistant who ensured that each group had a single response, and nothing awkward was entered into the document. It was explained to the students that this shared document was their notes for the course.

During the rest of the course, on an average, 12 questions were given for each three-hour session. At the end of each session, the online, shared document that was generated by the students was briefly brought on the screen, and discussed, and this was also repeated briefly at the start of the next session.

During the course, short assignments were given that corresponded to questions discussed over the four sessions.

Behind the scene activity (what students did not see)

The academic had put in significant effort in planning for the lecture, and the wording and order of each question was determined after significant searching on the web by himself and the teaching assistant. Further, the academic had prepared around 18 questions for the three-hour session, and dynamically selected the appropriate 12 based on the responses by the students.

Results and Discussion

The student feedback was outstanding. From the course being one of the worst rated in the School, it rose to be one of the outstanding ones. More importantly, the students attended more sessions than what was mandated, and most stayed for the entire sessions, not just for the attendance period.

During the sessions, all the students did attend to their personal emails, but no one was watching a soccer game, or a movie. One of the most positive outcomes was that by the end of the course, the students had begun to make friends among their colleagues and were feeling more confident with their interpersonal communication. It was also noted that their assignments were significantly improved compared with the previous years, suggesting that they had achieved the outcomes of the course.

A downside of this style was that it required full attention of a senior academic and a teaching assistant. Another downside was that organising such a course required significantly more planning and preparation before the actual delivery for both the academic and the assistant. While the PowerPoint style presentation provides the information to both the students and the presenting academic, this is not available to an academic teaching in this style who needs to be fully prepared with all details and must run the course from memory.

To conclude, Question Based Learning (QBL) was effective in engaging with students who worked in small groups to search for answers and report their findings on a common online document. While the students did regularly access their emails and messaging services, the final outcomes were excellent. However, this activity required a significant amount of preparation, attention to detail, and the extra cost of a teaching assistant. Also, this approach may not be suitable for very large classes.

References

Beland, L. P., & Murphy, R. (2016). Ill communication: Technology, distraction and student performance. *Labour Economics, 41*, 61–76.

Kumar, D. K., & Radcliffe, P. J. (2019). Teaching while surrounded by smartphones. I*n 2019 41st Annual International Conference of the IEEE Engineering in Medicine and Biology Society (EMBC), IEEE*. (pp. 4194–4197).

Chapter 7
Successful Strategies

The case studies and discussion in previous chapters have illuminated the successes and failures of attempting to ignore or use smartphones in an education environment (Kumar 2017). This chapter consolidates this material and provides a smorgasbord of techniques and tools that can be applied to your education environment. As already pointed out, it is essential to analyse the nature of your students as this will affect what strategies are liable to be successful. The table provides some of the strategies from this research, but you may find your own unique solution. The intent is not to give you a recipe, but to start a dialogue and help academics be aware of the problem, and their specific situation.

Reality Check

The first step is to do a reality check. It is essential to determine the relative competence and attitudes of the students in the classroom. When a culture change is needed, then careful explanations and incentives to change are required. If the students are medium achievers, then work with strategies that are suitable for them. Also, understand the support your University offers the academics; in terms of resources, and understanding. Also take into account the energy, passion and experience of any involved academics and tutors. It is also important that you take a full appraisal of the topic and its suitability for the planned style of teaching. And, finally, it is essential to realise that when you do make a significant change to the classroom teaching, there can be problems that you never thought about, and you have to be prepared for things to go wrong. You should plan how to monitor the progress of your new ideas and be prepared to change things as required when it becomes clear that something is not working.

Start the dialog

People are usually more accepting of changes, new methods, and rules when these have been communicated with them in advance. It is important that the reasons for making the change to the teaching methods, and the new rules that may be required, have been clearly explained at the start of the activity. For example, if you wish to

D. K. Kumar and P. Radcliffe, *Teaching Surrounded by Smart Phones*,
SpringerBriefs in Applied Sciences and Technology,
https://doi.org/10.1007/978-981-15-1401-2_7

ban smartphones in the class, it is essential that you explicitly tell students of this new rule, the method you would use to monitor the rule, and the consequences if this rule is broken. It is also essential that your colleagues and other relevant academics or managers, and if required, the peak student body, are fully aware of the rule.

If the rule is carefully explained to students and given reasons to which they can relate, and that it is in their self-interest or to their advantage, then they may take ownership and accept the rule with less resistance. Without such an understanding, the student body will try to undermine the rule and find ways to work around it. The rule and the reasoning may need to be mentioned several times in order the get the message across. While not all will come on board, the majority will, and the others will fall in place.

Explain consequences

It is essential that when explaining the rules, the consequences are made clear and then followed through when rules are violated. A failure to follow through on consequences will earn the wrath of students who would have been doing the right thing. Consider the rule "no phones in lab exams". There must also be a consequences statement such as "automatic zero for violating this rule". All detections must result in a mark of zero. In many cases there are school or university policies about what can and cannot be done and so rules and consequences must fit in with such policies. It is essential that you have the support of the institution, and this is discussed in advance with colleagues to ensure that you really do have that support. Without such support, it will become impossible to follow through on rules that may have significant consequences. If relevant, discuss these issues with the student bodies such as Union, to ensure they will not object to your plans.

Consequences can be individualised to the specific student or may be for a group where the group has misbehaved. However, it is essential that you have the resources to monitor, and the support to follow-through with the threat of punishment. The absence of this would lead to chaos and the resultant outcomes will be far worse than not having made any effort.

Strategy Table

This section consolidates and tabulates the strategies for teaching when surrounded by smartphones. It identifies the type of students and classroom conditions that may benefit from the strategy. The technology continues to evolve, and hence this table should be considered as the starting point for the academics.

Suitable student groups and classroom conditions	Strategy	Issues
All students	Encourage self-management	Requires strong motivational skills to get students to become self-managing
All students	Smartphone use monitoring	Must install (self) monitoring software. Student must regularly review phone use, possibly in alliance with parent or teachers
All students	Near term consequences of poor smartphone use	Must develop quizzes or other evaluation activities. Students must be aware of evaluation and consequences of poor effort
All students	Interactive lectures	Need tools such as Kahoot. Be wary of free text replies
All students	Keep students busy with tasks	A lot of work for staff. Lose slow students, bore fast students
High performing and motivated students. Most students in high-end courses such as medicine fall into this group. Some cultures where students tend to be more obedient may also come in this category	Do nothing	There is no serious problem. Well explained warnings about smartphones may be useful
Small classroom with availability of teaching assistant, and University supports the strict rules set by academics. Typically, the Universities are not dependent on the student fees, and student entry to the University is highly cherished	Ban the presence of smartphones and other Internet devices in the class	The academics have the resources and the University backing to ensure that the students do not have their smartphones with them during the class Only possible in some cultures where students are likely to be obedient, the class size is small, and there is a fear factor for consequences of misbehaviour
High-achieving students in small tutorial batches. Typical tutorials rather than lectures	Integrate the use of smartphones in the teaching by providing online material	The academics run the tutorials by providing the questions online. The questions need to be very well developed such that student activity can be monitored

(continued)

(continued)

Suitable student groups and classroom conditions	Strategy	Issues
High-achieving or post-graduate students in medium size classrooms (< 100). Availability of teaching assistants Sessions of 2 h or more	Use QBL (Question Based Learning) Convert the teaching material to small questions (typically 20–30 min) The students work in teams of 4–6 and find the material online. The academic discusses the question while joining each group for a short period The group enters their response to the questions by posting them on shared documents using methods such as Google-drive or One-drive, or similar	This method requires a very experienced and energetic academic who is got the energy to run these sessions, which tend to become very vigorous The teaching assistant has to monitor the entries on the shared drive to ensure there is no undesired remark or activity The final outcome of the shared activity becomes the resource of the entire class, of which the students take ownership Another significant outcome of this approach is that it develops the teamwork and communication skills of the students
Small groups where students will respond to teacher requests	Phones on, phones off. Phones may be used when requested by the teacher, then put away as requested by the teacher	Tardy students who do not put away their phone will hold up the class. The teacher must provide small set tasks through any lesson
Small groups where students will respond to teacher requests	Smartphones as a reward. Students can use their smartphone after completing work	The teacher needs to supply targeted activities which benefit the student's education
Students can work in groups toward a common goal	Collaboration tools are required to complete a task. There must be some advantage over doing the task with paper and pencil	All the usual methods and danger of group work apply here
Any student group that can perform work on a smartphone	Managed education tools that can guide and monitor student progress	These are complex tools that can be costly. Customizing content can be a lot of work. Students' progress at their own rate

Bibliography

Android accessibility overview—Android Accessibility Help. (n.d.). Retrieved July 10, 2019, from https://support.google.com/accessibility/android/answer/6006564?hl=en.

Carr, N. (2011). *The shallows: What the Internet is doing to our brains*. WW Norton & Company.

France bans smartphones in school—The Washington Post. (n.d.). Retrieved June 24, 2019, from https://www.washingtonpost.com/technology/2018/07/31/france-bans-smartphones-school/?noredirect=on&utm_term=.12f04620af74.

Irvine, D., Zemke, A., Pusateri, G., Gerlach, L., Chun, R., & Jay, W. M. (2014). Tablet and smartphone accessibility features in the low vision rehabilitation. *Neuro-Ophthalmology, 38*(2), 53–59. https://doi.org/10.3109/01658107.2013.874448.

Kumar, D. (2017). Research methods for successful PhD. Retrieved from https://riverpublishers.com/book_details.php?book_id=466.

Lin, Y. H., Lin, Y. C., Lee, Y. H., Lin, P. H., Lin, S. H., Chang, L. R., et al. (2015). Time distortion associated with smartphone addiction: Identifying smartphone addiction via a mobile application (App). *Journal of psychiatric research*, *65*, 139–145.

Mills, M. P. (n.d.). The mobile revolution has only just begun. Retrieved July 10, 2019, from Forbes website: https://www.forbes.com/sites/markpmills/2015/01/19/the-mobile-revolution-has-only-just-begun/.

Romaniuk, S. N., & Burgers, T. (n.d.). How the US military is using "violent, chaotic, beautiful" video games to train soldiers. Retrieved July 4, 2019, from The Conversation website: http://theconversation.com/how-the-us-military-is-using-violent-chaotic-beautiful-video-games-to-train-soldiers-73826.

Search Engine Optimization (SEO) Starter Guide—Search Console Help. (n.d.). Retrieved July 10, 2019, from https://support.google.com/webmasters/answer/7451184?hl=en.

The 20 Most Popular Android Apps in the Google Play Store. (n.d.). Retrieved July 28, 2019, from https://www.makeuseof.com/tag/most-popular-android-apps/.

Tobin, A. (n.d.). These were the most downloaded and profitable games and apps of 2018. Retrieved July 28, 2019, from Forbes website: https://www.forbes.com/sites/annatobin/2018/12/21/these-were-the-most-downloaded-and-profitable-games-and-apps-of-2018/.

The manufacturer's authorised representative in the EU is Springer Nature Customer Service Centre GmbH, Europaplatz 3, 69115 Heidelberg, Germany. If you have any concerns regarding our products, please contact ProductSafety@springernature.com

Printed and bound by CPI Group (UK) Ltd, Croydon, CR0 4YY

25/09/2026

02222891-0013